Chasing Giants

CHASING

IN SEARCH OF THE WORLD'S LARGEST FRESHWATER FISH

GIANTS

ZEB HOGAN & STEFAN LOVGREN

UNIVERSITY OF NEVADA PRESS | *Reno & Las Vegas*

University of Nevada Press | Reno, Nevada 89557 USA
www.unpress.nevada.edu
Manufactured in the United States of America
FIRST PRINTING
Frontispiece photograph by Brant Allen
Jacket design by Diane McIntosh

LIBRARY OF CONGRESS CATALOGING-IN-PUBLICATION DATA
Names: Hogan, Zeb, 1973–author. | Lovgren, Stefan, 1969–author.
Title: Chasing giants : in search of the world's largest freshwater fish /Zeb Hogan and Stefan Lovgren.
Description: Reno ; Las Vegas : University of Nevada Press, [2023] |
Includes index.
Summary: "Beneath the surface of the world's rivers and lakes swim mysterious giants, the real-life
    Loch Ness monsters and Bigfoots of the aquatic world. They are a diverse assemblage of poorly
    understood creatures, from gargantuan gars to sumo-sized stingrays. These ancient fish—some
    who have been around for hundreds of millions of years—play critical roles in their freshwater
    ecologies. Threatened by overfishing, habitat loss, dams, pollution, and climate change, the
    majority of the world's freshwater megafish are today at risk of extinction. As an aquatic ecologist
    and host of the National Geographic Channel's "Monster Fish" series, Zeb Hogan has spent more
    than a decade searching for and studying these increasingly endangered river titans. In this book,
    he teams up with award-winning journalist Stefan Lovgren to tell for the first time the remarkable
    and troubling stories of the world's megafish and chronicles the race against the clock to find and
    protect these ancient leviathans before they disappear forever."—Provided by publisher.
Identifiers: LCCN 2022029957 | ISBN 9781647790578 (cloth) | ISBN 9781647790585 (ebook)
Subjects: LCSH: Freshwater fishes—Ecology. | Freshwater fishes—Conservation. | Rare fishes—Ecology.
    | Rare fishes—Conservation. | Endangered species—Conservation.
Classification: LCC QL624 .H63 2023 | DDC 597.176—dc23/eng/20220713
LC record available at https://lccn.loc.gov/2022029957

The paper used in this book meets the requirements of American National Standard for Information
Sciences—Permanence of Paper for Printed Library Materials, ANSI/NISO Z39.48–1992 (R2002).

Contents

Photographs follow page 162.

Chasing Giants

Truly a Whopper

On May 1, 2005, Thirayuth Panthayom woke up, as always, before dawn. Humidity choked the air as he emerged from his bamboo lean-to hut on the banks of the Mekong River. It was the hottest time of the year in Hat Khrai, the small fishing village in northern Thailand where the twenty-nine-year-old student spent every fishing season. The monsoon period was coming; already the afternoon rains had grown stronger, causing the river to surge. Thirayuth knew what it meant. "Naam khun hai riip tak," they said—when the water rises, rush to catch fish. It was the one time of year to catch the Mekong giant catfish, the fish that Thai called *pla beuk,* or "buffalo fish."

Among the Thai fishermen, no fish inspired greater respect than the Mekong giant catfish. Stories of a fish that weighed more than three baby elephants put together seemed too incredible to be true. But Thirayuth knew they were no tall tales. He had grown up watching his father and other fishers in the village pull dozens of giant catfish out of the Mekong every fishing season. He himself had caught some. But three years had passed since anyone had caught a giant catfish. It was clear their numbers had declined dramatically. Some people wondered if there were any *pla beuk* left to catch.

Revered in Thai folklore, the giant catfish was believed to bring wisdom and good luck to whoever ate it. Special rituals had to be observed before fishing it. So before going out on the river that day, Thirayuth first went to the village shrine of the God of Catfish. Standing before it, he pleaded: "Please, Miss Boat, let me catch something today and I will sacrifice a chicken for you." Then he and four crew members got into the long boat that belonged to Thirayuth's father. Together they ventured out onto the chocolate-brown waters of the Mekong, to the site where they would place the mong-lai, a hundred-meter-long thick-twine floating gill net specially designed to catch giant catfish.

They had spent less than fifteen minutes on the water when suddenly a monster announced itself with four huge whacks of its tail—*Bam! Bam! Bam! Bam!* The fishers could immediately tell what was thrashing in the net: it was a *pla beuk;* no other fish grew as big, and this one was enormous. The creature was so big it would flip or sink the vessel if they tried to lift it into the boat. They would have to leave it in the water and drag both net and beast behind the boat back to shore.

It took more than an hour to reach land. By then, the brute force of the creature had dissipated. When one of the men threaded a sisal rope through its mouth and gills so that it could be tied securely to a bamboo pole stuck in the water, the toothless giant, bruised and scarred, put up little resistance. It took ten men to lift the giant fish out of the water and onto a tarp. It was then that the animal's gargantuan proportions were truly comprehended. The fish measured almost nine feet from head to tail. Someone suggested they should try to weigh it, but the idea was quickly abandoned. No scale, they all agreed, would be strong enough to hold such a leviathan.

The fish was a female. As required by their permit to fish for a critically endangered species, the fishers had to allow officials from Thailand's Department of Fisheries to harvest the female's eggs for the department's captive breeding program. Few giant catfish had ever survived such a harvesting process, in which hormone injections are administered and the animal's belly poked and prodded to extract the eggs. This time would be no different. The officials managed to remove some eggs at the riverbank, but they then decided to transport the fish, barely alive, to their research station at nearby Nong Luang for further extractions. It was a surprise to no one when the giant creature was later returned dead.

At that point, the fish was going to be cut up into steaks and sold in the market, where it might fetch up to two thousand dollars—a fortune in rural Thailand. But first the fishers decided to split the animal into large pieces and weigh those pieces individually. Slowly they added the numbers together.

The total came to 293 kilos, or 646 pounds.

~~~

I was in Mongolia, working on a conservation project focused on *Hucho taimen,* the world's largest trout, when I heard about the catch. I immediately got on a plane to Thailand. In Hat Khrai, I checked the local fishing association's weight records of all giant catfish that had been caught going back more than twenty years. The records confirmed that this was indeed the largest fish that had been caught in that area during that period, ten pounds heavier than the previous record holder.

There was no reason to doubt the account or weight accuracy. No rationale existed for the fishers to lie. They had no idea what the record was. They were just looking for the biggest fish they could catch and sell, to make the most money. If this particular fish happened to be ten pounds heavier than another, it meant an extra hundred dollars in their pockets, nothing more.

It was different for me. I had become absorbed by the question: What is the world's largest freshwater fish species? There was no obvious answer to this. Everyone knew that the African elephant was the world's largest land animal; that the blue whale was the biggest animal anywhere; and that the whale shark was the world's largest marine fish. But the identity of the largest freshwater fish species on the planet had still not been settled. How could this be?

And so I had an idea. With the help of the World Wildlife Fund and National Geographic, I put together a press release about the catch of the giant catfish. The headline asked if this was the biggest freshwater fish ever caught.

The story caught on right away. Media around the world reported that a 646-pound leviathan had been caught in Thailand, the largest freshwater fish ever recorded. I expected to be inundated with phone calls, emails, and photos from people who claimed to know of bigger catches. To my surprise, I got nothing.

As I continued to dig for answers, it became clear that no one had looked at this issue on a global scale. Here was a type of charismatic megafauna that in many ways had been totally unexplored. It wasn't just the Mekong River that harbored these giants. On rivers and lakes around the world, from the Amazon to Australia, there were untold stories of gargantuan gars and sumo-sized stingrays. Why did these fish grow so big? What evolutionary contingencies allowed giants that were so out of proportion to the freshwater fish we most readily summon to mind? Could there be other giant species that we didn't yet know about?

There was only one way to find out.

The Megafishes Project

The rainy season had just begun when I arrived in Chiang Mai, in the mountainous northern part of Thailand, in early September of 1996. The contrast with my home state, dry-as-bone Arizona, could hardly be more extreme. There, we were lucky to get a downpour every few months. Here, it rained every day and with such force that it felt like being doused by sheets of water rather than a rain shower.

I had never been to Asia. Everything seemed so vivid and alive. I felt as though I was walking around inside a personal movie theater where strange scenes unspooled with remarkable regularity. On my first day, I walked across the Chiang Mai University campus, hundreds of miles from the sea, when a purple crab the size of my fist scuttled sideways across the path. No one but me paid it any attention.

It had all come together so fast. One day I had seen an advertisement on a notice board at the University of Arizona, where I was a biology major. Now I was here, as a Fulbright scholar. I had no idea what to expect, beyond the fact that I was going to stay for a year, taking classes and conducting field research, because those were the terms of my scholarship. I didn't know much about the river I had come to study, the Mekong, which runs through six countries in Southeast Asia, including Thailand. I certainly had no idea that it was a river that would also come to run through my life.

My project was to research the impact that a series of planned dams along the Mekong, mostly in neighboring Laos, would have on migratory fish throughout the river system. That, at least, was an issue with which I was familiar. As an undergraduate student, I had done fieldwork on the Colorado River, where large hydro projects like the Hoover and Glen Canyon Dams had completely disrupted the lives of many native fish species, causing them to disappear.

The fear was that building dams on the Mekong mainstem, which was still free-flowing below China, where it originates, would have similarly destructive consequences, but on a much bigger scale. At the time of my arrival, more than two million tons of fish were being pulled out of the Mekong each year, making it the most productive river system in the world and feeding tens of millions of people living in the basin. Between 40 and 70 percent of the commercially important fish in the Mekong were thought to be migratory, and building dams in their migratory paths would severely restrict their movements. The problem was, no one knew how far certain species needed to go for such migrations, or whether the fish would be able to survive between dams. That's what I had come to find out.

To complicate things, Chiang Mai was a minimum six-hour drive from the Mekong River. I was worried that I would be stuck in a classroom, but my supervisor urged me to focus on my research interests and, when needed, skip classes. He advised me to take Thai classes and, as soon as I felt reasonably comfortable with the language, head into the field. "Figure things out for yourself," he said. It was good advice. I liked doing my own thing.

Figuring things out started with my host city, Chiang Mai, Thailand's second-largest city after the capital, Bangkok. It was a place easy to like. Near my apartment, which was down the street from a quiet forest temple called Wat Umong, was a market that sold anything that could be fried or cooked or otherwise prepared for food—from frogs, snakeheads, and buckets of mussels and snails, to big piles of greens and chilis, and fruits of all sizes, colors, and qualities. I went there for breakfast every morning and had strong Thai coffee with condensed milk, fresh-squeezed orange juice, and fried Chinese donuts. In the evenings, my classmates—all of them from South or Southeast Asia—would organize casual parties with simple food and impromptu group singing.

My first three months in Chiang Mai, I spent four hours a day taking Thai language classes. I spoke rudimentary French and Spanish, but Thai was tough because it was the first tonal language I had tried to learn. Words changed meaning if spoken in different tones, like pitches in music. Fortunately, I didn't have the stress of mastering the language for my university courses, which were taught in English, or to communicate with my classmates, because they almost all came from places outside of Thailand and did not speak Thai either.

By the end of the semester, I was able to have basic conversations in Thai. Armed with my new abilities, I decided not to take any regular classes the following semester and instead get into the field to do research.

I had read a few scientific papers suggesting that fish markets were one of the best sources of information on migratory fish, especially when it came to seasonal diversity—that is, which fish were found where and at what time of year. Every town in Thailand has a market called a *talat,* including a morning fish market, that often serves as the community's economic hub. Fish has played an essential role in Thai culture, as it has in all Southeast Asian countries. One of the earliest inscriptions in Thailand, found on an ancient stele, describes the thirteenth-century kingdom of Sukhothai as thriving at the time because "there is fish in the water and rice in the fields."

I picked four markets along the Mekong River to monitor over a two-month period. To get the best data, these locations needed to be at some distance from each other. My southernmost market, in the town of Khong Chiam, was an eighteen-hour bus ride away from my northernmost market in Chiang Khong. Fortunately, I had received a $2,000 grant from the Wildlife Conservation Society to do the monitoring, and with that money I was able to hire four Thai students to help. They lived in the four different towns full-time to collect data, while I circulated between the different locations.

And so I began to travel, up and down the Mekong. Bus travel was cheap in Thailand; the equivalent of ten dollars would get you around the whole country. But it was at times terrifying, especially on the overnight buses, with drivers speeding around sharp, blind corners in the middle of the night. I don't know how many times I was jolted awake at 3:00 a.m. as the bus driver slammed his brakes to avoid crashing into oncoming traffic. Simply hearing the name Nakorn Chai Air, the bus company I used, still makes me shudder.

The fish markets were a cacophony of sights, sounds, colors, and smells. In the early morning, the sellers—all women—put out the fresh catches that the fishers—often their husbands—had made overnight, and the grounds and bamboo stalls soon filled with shouting and crowding and a dizzying array of bizarre-looking fish.

Species like giant snakeheads—massive, iridescent, purple and green fish that got their name because of their serpentine heads—were piled next to Siamese mud carp, gourami, swamp eels, climbing perch, redtail loach, and tinfoil barb. After all, the Mekong River is home to nearly a thousand different species of fish, making it the most biodiverse river in the world after the Amazon and Congo.

One of the most common sights in the markets were *Siluriformes,* more commonly known as catfish. Catfish are ray-finned and scaleless, named for their prominent barbels, which look like a cat's whiskers. There are thousands of different species of catfish around the world, ranging in appearance, size, and behavior. In fact, one out of every twenty vertebrate species on the planet is a catfish.

I began to focus on one family of catfish in particular: the pangasiids, or shark catfishes, of which there were about a dozen described species in the Mekong basin. I saw eight to ten of them regularly in the markets. Among these were aptly named fishes like the striped catfish, the mouse-faced catfish, the snail-eating catfish, and the gargantuan dog-eating catfish, which, according to legend, fishermen used to catch with hooks baited with chunks of dog.

What the pangasiids had in common was their commercial and cultural importance. In most cases they were abundant, which made them perfect study animals. Key to my research, they appeared to be highly migratory, making transboundary journeys up and down the Mekong. They were a common currency that was recognizable and had value in all the Mekong countries.

However, it was difficult to tell them apart. Many of the pangasiid species found in the markets were about the same size, and they had similar colorations and shapes. But there was a way of identifying them: by their teeth. At first glance, they didn't appear to have teeth, more like pads of different shapes. When you looked closely, however, some of those padlike teeth were shaped like a crescent moon, others like the end of a hockey stick or the bottom of a dog's paw. Gradually, I began to recognize the patterns, like words in a foreign language.

As we collected more data, I started piecing together the migratory patterns of the pangasiids. They were seasonal spawners, grouping together for a few months starting in May to breed later at the beginning of the rainy season. While larger species, like the dog-eating catfish, could be found in the northern section of the river between Thailand and Laos, and smaller species inhabited the middle stretches of the Mekong, it was clear they all made substantial migrations to complete their life cycles. The proposed dams, it seemed obvious, would block those migration routes and threaten the very existence of these important fish.

The one pangasiid species that I never saw in the markets was *Pangasianodon gigas,* the fish that the Thais called *pla beuk* and which went by the common English name of "Mekong giant catfish." I had heard the accounts of its enormous size, but catches of the giant catfish had dwindled in recent years, judging by its absence from the markets. Yet only a few people were studying this giant or raising calls about its decline.

One was a young Thai biologist named Chainarong Sretthachau. He had written an excellent report about the endangered status of the Mekong giant catfish. When a mutual friend told him that I was interested in seeing the pla beuk, Chainarong called me to ask if I wanted to go with him to the province of Chiang Khong to see if we could find any giant catfish.

This was in May, during the annual pla beuk fishing season, and the chances of seeing a giant catfish, I was told, were pretty good. In the early 1990s, fishers in Chiang Khong regularly caught dozens of giant catfish over a relatively short fishing season. So I said yes, and together we drove for six hours to Chiang Khong, into the area known as the Golden Triangle, where northern Thailand, Laos, and Myanmar meet, finally arriving in the fishing village of Hat Khrai.

We arrived just when a local fisher had caught a pla beuk. The giant had been caught in a gill net placed in a shallow river channel of the Mekong. After bringing it ashore, the fisher had tied its mouth and tail to a thick bamboo pole by the river's edge. It was there that we found it, still alive.

The largest freshwater fishes I had seen in the wild prior to coming to Thailand were thirty-pound carp and catfish on the Colorado River. I had seen big fish in the markets in Thailand, like the dog-eating catfish, but they had all been dead. This was different. I wasn't mentally prepared to encounter an animal of this size—eight feet long, as big as a bear—in the wild. There was something both wonderful and unsettling about it. The idea of a fish this huge living in fresh water was completely counterintuitive, as though I had misunderstood something fundamental about rivers and the animals that lived in them.

I was lucky to see such a rare fish, but my mood soon turned grim. With its scaleless surface, broad head, large, low-set eyes, and downturned mouth, the Mekong giant catfish gave off a sorrowful impression. It was a gentle creature, persisting on an herbivore's diet of plants and algae. As I stood there watching it struggle against the ropes, a wave of sadness washed over me because I knew it was going to die.

I returned to the United States, and in the fall of 1997 began a PhD program at the University of California, Davis. My thesis was going to be about migratory catfish in the Mekong River. I knew this meant spending a lot more time in Southeast Asia over the next few years. I would not be able to do it from the confines of a California classroom.

I was keen to find other people to collaborate with. During my year in Thailand I had met a Canadian man named Ian Baird, who was living and working in southern Laos. A few years older than me, Ian had dropped out of college in the mid-1980s and moved to Thailand, where he ended up working with coastal fisheries. He became interested in dolphins and, in 1993, moved to Laos after getting approval from its Communist government to start a program called the Lao Community Fisheries and Dolphin Protection Project, which operated out of a small fishing village called Hang Khone, near the border with Cambodia.

The project was ostensibly about the tiny population of rare Irrawaddy river dolphins found in the area, but Ian's main focus was researching and implementing community-based fisheries management and working with local fishers. Among the most common fish caught in the area were two pangasiids that were of great interest to me. So I asked Ian if I could come and stay with him in the village for a while to learn more about the fish. He agreed.

Hang Khone is situated on one of four thousand islands that dot the Mekong in the southernmost province of Laos, just downstream from a series of cascading waterfalls known as the Khone Falls. The most beautiful spot on the entire Mekong, the Khone Falls were the Mekong's only waterfalls and, unlike the rest of the murky monster that is the Mekong, during the dry season the water there was clear and accessible, perfect for snorkeling and seeing thousands of fish.

The village itself was equally picturesque. Forty-five families lived in traditional Lao wooden houses with vegetable gardens, surrounded by coconut trees and bright, green fields with small stupas in the middle of them. The house where I stayed was the only one with a generator, and after dark two dozen people gathered at the house to watch TV together. Everyone was very friendly.

And everyone fished, often using artisanal techniques that had been developed over generations. They employed bamboo filter traps, fencing, wicker baskets, and nets hung from poles. In many places, elaborate scaffolding had been constructed across river channels so that fishers could get closer to the fish. In other places, the fishers had to move across the falls by balancing on thick-gauge wire or scampering across slippery rocks. Accidents were common.

Ian, who called himself an ecological Marxist, had helped devise a system that encouraged the fishers to pool their resources and share fishing yields equally. His philosophy was that the local fishers knew best how to manage their own fisheries. He also considered them the best source of information on what the fish were doing in the river. This I came to believe too.

Researchers often dismissed fishers' accounts as unreliable, but those who fished for work were the ones who knew what fish were being caught where and when, and in most cases they didn't have any reason to provide misleading information. The trick was documenting the fishers' knowledge and fish catch in a way that was scientifically accurate and could be integrated into a broader body of scientific work on the topic. I never would have been able to do my research in Hang Khone without the collaboration of its fishers.

One of the two catfishes that I was focused on was *Pangasius macronema,* a small but abundant species that didn't have an English common name. Around the Khone Falls, this fish was caught in large numbers in April and May, at the end of the dry season. At that time, the water was low and the fish had only one passable channel to use, which meant the fishers could intercept them. But now there were discussions about building a dam on that very channel. This would block the migration of not only the *P. macronema* but also many other migratory fish with great economic and social importance.

The other catfish was the *Pangasius krempfi,* or silver-toned catfish, which could grow more than three feet long. This was an interesting species, because it seemed to be anadromous—that is, it moved from the sea into the river. Ian called it a "Mekong salmon." He had seen the fish on the coast of Thailand, and I had seen it in Nong Khai, in northern Thailand, about a thousand miles upstream of the river estuary. And in May and June, it was caught at the Khone Falls. Putting the pieces together, there was good reason to believe the silver-toned catfish made very long migrations up the river to spawn.

But believing is not knowing. No matter what we thought we knew, it was just a hypothesis. Just because the same species was found in far-apart locations didn't mean the fish were part of the same population. We needed to prove it, and I had an idea of how to do that.

I had met a scientist at the University of Hawaii who studied the migratory behavior of a fish called gobies by using a novel technique of analyzing the fish's otoliths, or ear stones. The otoliths could be used to tell how old a specimen was. But the researchers had also found that an analysis of the ratio of calcium to strontium—an alkaline earth metal that is prevalent in spices and whole grains—in the ear stone could tell you whether a fish had

been living in salt water or fresh water, because strontium concentrations in the ocean were far greater than in rivers or streams.

So we began collecting the silver-toned catfish. Over two years, thirty-six specimens were sent to Hawaii for analysis. Lo and behold, the results showed that the otoliths of the fish contained significant amounts of strontium, which was clear evidence they had lived in salt water.

But sometimes science has a way of muddying things up. Further analysis of a larger data set showed that the silver-toned catfish indeed spent time in both salt water and fresh water, but also sometimes in brackish water. It seemed like they were roaming around more opportunistically than we had initially thought, and only eventually migrated to fresh water to spawn.

It was an important lesson: what we think is happening might not be exactly what is happening, which, of course, is why we do research. One study builds on another, which builds on another. It did not change the basic conclusion. Ian was right: *P. krempfi* was a "Mekong salmon" that made incredible migrations from the ocean and far up the Mekong River. Later, when we published our results, it was the first documented case of anadromy in a Mekong River species.

For many years, the plans for the dam at Hou Sahong, the river channel so crucial to fish migration, seemed to have stalled in the face of widespread protests from environmentalists and local activists. But eventually those plans forged ahead. In 2016, construction of the dam began and the fishery that had existed there was destroyed.

~~~~

From the Khone Falls, the Mekong drops into Cambodia—the ecological heart of the Mekong River basin and the epicenter for fish. Cambodia seemed to be the best place to find answers to many of the questions I had about the migratory habits of Mekong catfish. But my first encounter with Cambodia, when I traveled there in early 1997, had little to do with fish. Instead I got a crash course in Cambodia's turbulent political history, as the country was teetering on the brink of civil war.

Cambodia, a former French colony, had already endured decades of conflict, going back to the Vietnam War when it was bombed over several years by US forces targeting North Vietnamese bases inside Cambodia. In 1975, as the Vietnam War officially ended, the country was taken over by the murderous Khmer Rouge regime, led by madman Pol Pot, who undertook a radical experiment to create an agrarian utopia in Cambodia, with disastrous consequences. During his reign, an estimated two million people died of starvation, disease, overwork, and executions in one of the worst mass killings of the twentieth century.

The eventual ousting of the Khmer Rouge by Vietnamese forces in 1979 led to more than a decade of Vietnamese occupation of Cambodia. A peace accord in 1991 paved the way for free elections in 1993, which were won by royalists led by Prince Ranariddh. But Hun Sen, the ex-Communist leader of the Cambodian People's Party, refused to concede power, and a standoff ensued, which ended when the two men agreed to a power-sharing deal.

That alliance had just fallen apart when I arrived in Phnom Penh, the capital. Supporters of the two men clashed openly in the city. The fighting, which killed dozens, ended with Hun Sen ousting Ranariddh, who fled into exile in France, while Hun Sen remains prime minister of Cambodia to this day, making him the longest-serving prime minister in the world.

Even after the coup, Phnom Penh felt like a lawless Wild West, which appealed to a certain set of tourists, like those who liked to fire heavy-duty bazookas at special "shooting ranges." Restaurants downtown sold "happy pizzas," which were pizzas laden with globs of hash. The title of a book by Amit Gilboa—*Off the Rails in Phnom Penh: Into the Dark Heart of Guns, Girls, and Ganja*—sums up the dark edges of the expatriate lifestyle in Cambodia at the time.

But I was there for the fish. Nowhere in the world do freshwater fish play a more important role to humans than in Cambodia, where the average person consumes a staggering 115 pounds of fish per year. At certain times, trey riel, or "money fish," a small, shiny fish reminiscent of a silver dollar, is caught in Cambodia in volumes so massive that piles of it sitting on the riverbank reach ten feet high, like heaps of treasure.

One reason for this abundance is the annual flooding of the Mekong, which produces an extraordinary natural phenomenon involving Southeast Asia's largest lake, the Tonle Sap, which sits in the center of Cambodia. It begins with the monsoon rains, which usually start in May and run through October. Swelled by intense rainfall, the Mekong rises as much as forty feet, sweeping with it not only sediment essential to agriculture but also enormous amounts of larvae and tiny fish produced from adults spawning in upper portions of the river.

While many of these juvenile fish are dispersed throughout the floodplains of central and southern Cambodia and Vietnam, large numbers are diverted into the Tonle Sap River, a tributary of the Mekong that joins the larger river at Phnom Penh, the Cambodian capital, and connects it with the Tonle Sap Lake. During the dry season, the waters of the Tonle Sap River flow into the Mekong. But during the rainy period, the Mekong waters instead push up and into the Tonle Sap, forcing the tributary to reverse its course, making it one of the only rivers in the world to do so.

When this happens, the tiny fish move with the river into the lake, which in the dry season covers an area about the size of Rhode Island, but at the height of the rainy season expands four- or six- or even eight-fold—bigger than the state of New Jersey. The incoming waters flood surrounding forests, creating excellent habitat to feed and grow hundreds of species of fish, as well as many birds, reptiles, and amphibians. It's truly an ecological wonder, which is why UNESCO designated the Tonle Sap Lake an official biosphere reserve in 1997. It is also a critical source of food, with some studies suggesting that more fish are drawn from this one lake than from all of North America's rivers and lakes combined.

When the rains stop, the lake drains, and many of the fish, increased in size, are pushed into the Tonle Sap River, creating another massive migration that local fishers have taken advantage of for millennia by fixing all manner of traps and nets to snare the aquatic travelers as they make their way back to the Mekong. The fish that make it past this gauntlet may then move up the river to spawn. Then, when the rains eventually come, the whole process is repeated.

I wanted to know where exactly these fish swam. Where did they travel after leaving the Tonle Sap River? Upstream or downstream? How far did they go? These were all unanswered questions. There were a few people, like Tyson Roberts, a distinguished ichthyologist, who had written scientific papers about Mekong fish migrations and diversity, but no studies had been conducted on a larger scale.

One institution that came to facilitate my work was the Mekong River Commission (MRC). This organization had been set up in 1995 by four countries that shared the Mekong River—Cambodia, Laos, Thailand, and Vietnam—to jointly manage water issues. Its fisheries program was headquartered in Phnom Penh and housed on the ground floor of the Cambodian Department of Fisheries, in what had been the US Embassy before the Khmer Rouge takeover.

I spent a lot of time there. It was a dark and crowded office filled with young Cambodians eager to learn about fisheries. With local scientific knowledge virtually nonexistent, a main goal of the MRC was to build up technical expertise among the Cambodians. That responsibility was placed in part on the commission's technical director, a genial Dutchman named Niek van Zalinge, who was very supportive of my research. Niek suggested that, for access to fish, I should look at a particular fishery located in the lower part of the Tonle Sap River.

The fishery used dais, or bag nets, in rows next to each other in the river, about sixty in total. The funnel-shaped nets were huge—330 feet long and 80 feet in diameter at the mouth—and they functioned like stationary trawls: With their tops sitting just below the surface and their bases on the river bottom, they trapped whatever was moving downstream. The first row of four nets was just north of the city, and the final station was situated twenty miles upriver. The dai fishery ran from October to March, when the fish came out of the lake. During peak times, a single net could catch over five tons of "money fish" in one hour.

My main focus was the catfish. But this time I wasn't looking to record catches or analyze genetics for migratory clues. I was going to test something that had never been done in the Mekong: follow the fish around by fitting them with acoustic or radio transmitters. While biotelemetry systems were commonly used in North America and Europe to study fish movements, this high-tech strategy had never been applied in the Mekong River basin because most people believed the river system was too large and complex. But I was able to obtain a research grant to try it in the Mekong. The dai fishery was the perfect provider of fish to tag, because all kinds of fish were caught there, including many types of catfish.

One pangasiid species found in the Tonle Sap was the *Pangasius hypophthalmus,* also referred to as the iridescent shark catfish because of the shiny skin exhibited in juveniles. Primarily known as the river catfish, it could grow over four feet long and was heavily cultivated for food, especially in Thailand, where it was marketed under the name *swai,* and Vietnam. Juveniles were also sold as pets for home aquariums, though the river catfish was not an easy fish to keep because it required both space and company.

The river catfish appeared to be in decline, and I decided that it would be a good study animal for our biotelemetry work. With the help of the dai owners, we collected eleven river catfish and outfitted them with acoustic transmitters as well as plastic tags that carried instructions in the Khmer language to "Please return to the Department of Fisheries," and then we released the fish back into the river.

One evening, we were cruising on the Mekong about twelve miles from its confluence with the Tonle Sap when suddenly the hydrophone we were trailing from our survey boat gave off a ping, and then a few more. It was a clear contact. It indicated a fish that had moved out of the Tonle Sap River and was now making its way up the Mekong. Although we never actually saw the river catfish, we were able to identify it by the unique code programmed into its transmitter. It was a seventeen-kilogram specimen we had tagged ten days earlier. Two months later, the same fish gobbled up the baited hook of a local fisher's line almost two hundred miles upstream from Phnom Penh.

Fishers also captured tagged specimens in this same area, suggesting that the migration route—from the Tonle Sap Lake, down the Tonle Sap River and up the Mekong—was typical of river catfish, which would move into deep-water areas of the Mekong River to survive the dry season and then migrate upstream to spawn with the onset of the first heavy rains in May and June.

In the beginning, the Cambodian fisheries officials I worked with— Thach Phanara and Em Samy—and I traveled by moped from the city and parked among the makeshift shacks of the people living along the river. From there, the fishers took us to their fishing stations. Later, I bought my own small boat, which we drove from the city directly to the floating houses that were attached to each bag net. There we sat and made small talk about fishing and how the gear worked. I came to cherish this time spent with the dai fishery families. I made friends with several of the operators, even a mean lady called Srey Pheap, who caught more catfish than anyone else but used to hit me in the back for no reason.

My favorite fishing family was one headed by a happy, smiling man named Mout Sitha. He didn't speak English, but two of his three daughters— Sokha, Phea, and Phors—did. I asked Phors if she could write down some information about the fishery and how it had changed over the years. What I got back from her were pages and pages of notes that meticulously covered everything about "the operation at lot 2A at kilometer 6, Phnom Penh."

In the past, Phors wrote, the dai fishery was not as developed as today, though most dai fishers both then and now were of Vietnamese descent or Muslim. She explained that the floating houses had been introduced in 1965, and included facts about the different groups of fish that dai owners kept and fed underneath their houses: slimy fish, cruel fish, and scaled fish. The slimy fish liked to eat rotten food.

The notes included detailed descriptions of how to make nets, prepare rafts, and operate anchors, as well as how to cook prahok, the fish paste made from trey riel that was the staple ingredient for all Cambodian food. She had also listed expenses for running the dai operation, which included everything from diesel and nylon cable costs to cigarettes and medicine, and totaled $5,395.50 for five months. The report also said that big fish—and, specifically, the Mekong giant catfish—had been a lot more abundant in the past.

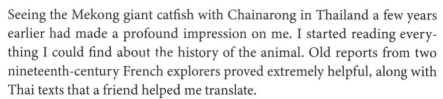

Seeing the Mekong giant catfish with Chainarong in Thailand a few years earlier had made a profound impression on me. I started reading everything I could find about the history of the animal. Old reports from two nineteenth-century French explorers proved extremely helpful, along with Thai texts that a friend helped me translate.

A century ago, the range of the Mekong giant catfish had spanned the entire length of the lower Mekong River and its tributaries. But in the 1930s and 1940s, the fish began disappearing from large segments of the river. There were estimates that the total number of giant catfish had decreased by as much as 90 percent during the past two decades. No one knew how many were left, but the number could be as low as a few hundred adults. Overfishing had been the primary culprit in the giant catfish's decline. Now the damming of the Mekong could deal a final blow to its survival.

After leaving Thailand, I had written a one-page paper called "The Quiet Demise of the Mekong Giant Catfish." It was a plea for the giant catfish. Few people knew or cared about this huge, iconic species that was possibly headed for extinction, and it needed advocates. I wanted to do what I could to understand and protect it.

I thought that Cambodia had become the last refuge for the Mekong giant catfish, but it was clearly declining there too. Culturally, the species was held in high esteem. Known as *trey reach,* or "royal fish," Mekong giant catfish were depicted in wall carvings found around the temples of the ancient Khmer empire that built Angkor Wat. However, unlike in Thailand, the fish was not considered a prized food catch among Cambodians, who simply didn't like the taste of it. So it sold for fifty cents a kilo here, less than a tenth of the price it fetched in Thailand.

I was surprised that the Mekong giant catfish was harvested and sold for so little money in Cambodia. To me, this was an opportunity to save these fish from getting killed. Together with Niek, I came up with an idea. What if we offered to buy the giant catfish from the fishers who had caught them, so that they could be released, alive, back into the river?

Similar compensation schemes had been used in many other places. In the United States and Canada, ranchers had sometimes been reimbursed for livestock killed by wolves or bears. In some African countries, farmers were paid to not harm elephants that encroach on their land. In India, the same was done with tigers. And yet it was considered to be a controversial practice. Did it really work? And what happened if you could no longer pay?

As far as I knew, it was not something that had been tried in Cambodia, where conservation work was still in its infancy. The country had a spotty environmental record. In Mondulkiri forest, once dubbed the "Serengeti of Asia," almost all of the area's abundant wildlife, including Indochinese tigers and Asian elephants, had been wiped out. If these big, iconic animals had been allowed to be eliminated, could I really hope to save a fish species that few people knew much about?

The capture or sale of the Mekong giant catfish was illegal in Cambodia, but that didn't mean it didn't happen. The rationale for my scheme was that it would prevent fishers from selling the fish without recording it. I didn't want the program to be seen as an incentive for fishers to catch endangered fish, but those catches were happening anyway. This way we would be able not only to save the fish, but also track it.

To make the program work, I needed buy-in. I had developed close relationships with the dai fishers. But would they actually call me if they caught a giant? What about other fishers on the river? Importantly, the plan got the approval of the Department of Fisheries, which agreed to make a formal request of cooperation to all fishers. In addition to a World Wildlife Fund grant, I received funding from National Geographic to pay for field expenses and fish purchases, as well as stipends for my staff. The funding allowed me to continue tagging fish as well. It was exciting to engage in this pioneering work. But I had no idea how successful it was going to be.

It didn't take long, however, before reports began pouring in from fishers up and down the Tonle Sap River who had caught fish that were on the conservation list: Mekong giant catfish, giant carp, and seven-striped barb. A jolt of trepidation hit me every time I got a call. There was Phanara on the other end, excitedly telling me in broken English that a giant had been caught and we needed to move immediately. When we arrived at the nets, there was inevitably a thick rope tied from a floating house. On the other end of it was a really big fish. This was the moment of truth. If the fish was swimming upright, with its gills opening and closing strongly, that was a good sign. If the fish was upside down, barely breathing, showing serious scrapes or cuts, my gut dropped because I knew it was probably going to die.

The fish all suffered from capture myopathy, or shock disease, a little-studied condition documented in wild animals, such as hares and birds, that had been captured or handled. The stress of capture leads to overexertion of the animal's muscles. When we saw how weak the fish were after capture, we constructed a tank and pumped oxygen and antibiotics into the water in an effort to speed the fish's recovery.

Large fish captured in nets had it worst. When caught, they were forced into the cod end of the net, which wasn't big enough for a giant catfish. It would get incredibly scraped up in there, and it would struggle to breathe. Once the fish had been removed from the net, the priority was to tag it and, as quickly as possible, release it back into the river. Time was of the essence.

But we had to be careful with where we released them, too. Fish released near the riverbank just floundered by the shore, and when we released them upstream of the fishing areas, they were recaptured in nets downstream. So we had to release the fish in the middle of the river and downstream; but this was no easy task. Phanara or I would drive the boat, while the other person stayed in the water to hold on to the fish as the boat pulled us out.

Our boat was so small and the fish so big, that we couldn't drive upstream or we would swamp the boat. We had to work our way slowly downstream, all the while keeping to the middle of the river where the water was deepest and the current strongest. The trick was not to fight the current. If you fought it, its tremendous force could thrust you into a piling or into the vortex of a massive eddy, where surface water was pulled into the deep.

Finally, I arrived at the middle of the river, with a several-hundred-pound catfish next to me. I held on to it by its mouth, pulling it back and forth, trying to get water flowing over its gills and to get its tail kicking again. Holding on to its mouth was possible because the giant catfish doesn't have any teeth. Its lips are hard and round, and gripping them is like holding on to a slippery rim of a basketball hoop. I stayed with the fish until I got a sense that it could swim on its own.

At that moment, I straddled the giant and pushed its head down, guiding it into the deep. And then I followed. The water darkened against my eyelids. My ears popped, but I kept going. How deep was I? Ten feet? Fifteen feet? I pushed the giant fish down and down until I could go no farther. That's when I let go, and if all went well, the giant kept going, into the depths of its underwater world, while I returned to mine.

~~~~~~

The following year I earned my PhD. One chapter of my thesis focused on the anadromous *P. krempfi*, with another chapter detailing the communal fishery at the Khone Falls. There was a short section of catfish population genetics, with the rest devoted to my work on the endangered, migratory fish of the Cambodian Mekong and the Tonle Sap. Although my work wasn't as technical as the theses of most of my fellow students, it was unusual enough to attract attention. The UN Convention on Migratory Species recognized the research with its "best thesis" award, and *American Scientist* published an abbreviated version of it as its May 2004 cover story titled "The Imperiled Giants of the Mekong."

By this time, my work was fully focused on the big fish. The giants of the Mekong were under all sorts of threats, and some species were perhaps on the brink of extinction. Yet no one seemed to know much about them.

It wasn't just the Mekong giant catfish, or the giant Siamese carp, or even the river catfish. There were others, like the seven-striped barb, a beautiful cyprinid that could grow up to 150 pounds and was once found throughout southern Asia. There was the massive goonch catfish, a real gremlin of the deep with a flattened head, fleshy barbels, and a mouth full of nail-like teeth. And the giant salmon carp, which was identified only in the early 1990s, yet hadn't been seen since 2000, as if it had come into existence only for a fleeting few years.

And then there was the giant freshwater stingray, a fish so large and powerful that, when hooked, had the capacity to pull a fisher and his boat up and down the river for hours. I had heard stories about the rays, and I had read accounts of their existence in Thai newspapers, but in 2002 I got to see one myself.

Niek told me about a large stingray that had been caught in Prey Veng, a southeastern Cambodian province. The fishers there were keeping it alive. Niek and I hopped into a four-wheel-drive and raced from Phnom Penh to Prey Veng in record time, thanks to the commission's Cambodian driver, a mild-mannered man named Mr. Kea, whose maniacal speeding and unabated honking I appreciated because the life of a rare fish was hanging in the balance.

The fishers were waiting for us on the far side in a spot where the river was wide and slow. A boat delivered us ashore, and we approached like detectives arriving at a crime scene. A rope tied to a house on stilts led into the murky water. One of the fishers pulled on the rope. After a few seconds a massive, dark shape breached the surface. First, its humped back emerged, then two beady eyes on top of a flat, disk-shaped body. The rope securing the fish had been threaded through the stingray's spiracles, the holes behind its eyes through which the rays breathe.

Witnessing the massive ray rise from the water was similar to seeing the Mekong giant catfish for the first time. The stingray was the size of a trampoline, seemingly too big to live in the confined space of a river, even if the river was huge. But at least the giant catfish looked like a fish, with fins and an elongated body. This beast was different. The stingray gave the impression of an otherworldly creature. Not only did it seem out of place in the river, it seemed out of place on Earth, a beautiful alien.

It was one of the biggest fish I had ever seen, measuring 4.13 meters, or 13.5 feet, in length, including the tail. We wanted to weigh the ray, too, but the fishers had no scale, and even if they had, it wouldn't have been big enough. I negotiated a twenty-five-dollar payment for the stingray, which seemed very cheap, and together with the fishers we released it back into the river. As we headed back to Phnom Penh, I wondered if I would ever see one again.

There was good reason to believe the giant freshwater stingray was rare, and maybe in trouble. Accounts from other rivers, most notably the Chao Phraya in Thailand, indicated declining populations of giant stingray and a particular sensitivity to water pollution. But all the big fish seemed to be in big trouble. Their huge size made them vulnerable to overfishing, because humans often preferred to hunt the largest animals. Huge fish required more food and more space, and space was in decreasing supply because humans also liked to build on the rivers where the fish swam. In most cases, the big fish were long-lived but slow to mature, and that was a problem too, since they had to survive a long time in order to reproduce.

I felt an urgency to take action to protect these animals. But first we had to increase our knowledge about their biology, their behavior, their very existence. What did the demise of the river giants say about the health of our freshwater ecosystems? That was a larger story, and maybe it could be told through the giant fish. Maybe they could be conservation flagship species—the equivalents of tigers and elephants—rather than freaks of the deep. Maybe they could represent the wonders of life in rivers and lakes. If only people knew about them.

To tell their story, I needed to venture beyond the Mekong River and Southeast Asia. I needed to know if what was happening to the megafish in the Mekong was happening in other parts of the world, and what that meant not just for the future of the big fish but for us humans as well, for the rivers and lakes on which we depend. I proposed the idea to the World Wildlife Fund and to National Geographic. They both jumped on board, effectively launching the Megafishes Project that would take me around the globe in search of our largest freshwater fishes.

Contenders and Pretenders

We had figured out the age of the universe and the distance to faraway stars. We understood the complex inner workings of cells and atoms and how to alter them to cure disease or generate energy. And yet the question of what was the world's largest freshwater fish remained unanswered. It seemed absurd.

I thought there must be a simple, definitive answer—one fish, living in fresh water, growing bigger than all others. Perhaps it was the Mekong giant catfish. Or maybe a catfish in South America or in Africa or somewhere else, a fish that was unknown to me but well known among anglers and local fishers.

I started with the Internet, searching with simple terms like "world's largest freshwater fish," "largest fish Amazon," and "largest freshwater fish Africa." After gathering pages of accounts of big fish, I realized there were many contenders for the title, with claims of record breakers on almost every continent with almost every type of fish imaginable. But it also became clear that many of the "records" were fabrications or exaggerations, like the oft-told accounts of scuba divers finding car-sized catfish below dams, or of man-eating fish in Asia or the Amazon.

In pursuit of more reliable information, I went to the university library and looked for peer-reviewed literature on record-breaking fish. These papers had the advantage of being fact-checked for accuracy. Even so, many accounts were hard to verify, sometimes because of the age of the report or because the author had not listed the maximum weight of the fish. Even FishBase, a generally reliable peer-reviewed online global database of fish species, had errors. It listed the maximum size of the wels catfish (*Silurus glanis*) as 306 kilos (675 pounds), but further digging showed that the weight was based on a completely unsubstantiated account.

It was clear that no definitive answer existed. More than that, no one had even really asked the question. Answering it was an opportunity to explore something that had not been documented before on a global scale. I had followed the adventures of National Geographic explorers like Brady Barr, who set out on a global search to find all living crocodilians, and Mike Fay, whose MegaTransect project, a 465-day walk through the Congo jungle, had helped put together an ecological map of one of the least studied areas of the world. This was a chance to embark on a similar quest.

However, it was not a typical academic project. Some of the colleagues I told about the project said that if I wanted an academic career, I should probably focus on more traditional research activities. Not only that, the project I was proposing—raising awareness about the ecological status of the world's rivers and slowing the extinction of Earth's largest freshwater fishes—seemed nonscholarly and too broad to some.

I saw it differently. For me, this was an opportunity to meld the academic rigor and resources available through my university with the outreach and educational strengths of National Geographic. As Mike Fay said of the MegaTransect project, "It's the golden combination, connecting science and adventure to reach millions of people."

I knew that to be successful, the project required both extensive thought and painstaking planning, starting with the establishment of clear parameters for the study: What, after all, qualified as a megafish? After much deliberation, I decided to cap the search to those fish species with the potential to grow at least six feet long or two hundred pounds, or both. I say potential, because I had a hunch that some species may have been able to grow bigger in the past and I wanted to be able to consider all evidence, even anecdotal. To be crowned the record breaker, though, the catch had to be verified, and I would go by weight, not length. There were some extremely long and skinny fish out there. But to me, the largest meant heaviest, just like the African elephant was considered the largest animal in the world, not the giraffe.

My first list of contenders numbered twenty-one species. Many were found in Asia, with a half-dozen species from the Mekong basin alone, plus giants like the Chinese paddlefish in the Yangtze River and the mighty mahseer in India. But megafish that fit the project criteria could be found on all continents, except for Antarctica. Weighing in from North America were living dinosaurs like the alligator gar and lake sturgeon; from South America, the amazing arapaima and goliath catfish. The freakishly large Nile perch from Africa qualified, as did the ferocious wels catfish that swam in European rivers. And of course there were giants in Australia too, like the Murray cod.

It was striking how diverse the contenders seemed. What did the giant freshwater stingray have in common with the Colorado pikeminnow? But there were many commonalities too. Most of these fish had not been studied well, presumably because they were found in remote and isolated areas. But perhaps most importantly, the fish—with few exceptions—all seemed to be in decline. On the list were at least a half-dozen giants deemed to be critically endangered.

One fish that I did not include on the list was the beluga sturgeon, which various news articles described as the world's largest freshwater fish species. Found primarily in the Caspian and Black Sea basins, the beluga could grow to immense proportions. A female caught in the Volga estuary in 1827 reportedly weighed 1,571 kilos, or 3,463 pounds, and measured 7.2 meters (more than 23 feet) in length. Such giants had become increasingly rare in recent decades, due to the heavy fishing of the species, but beluga sturgeons over 1,000 kilos could still be found.

But the beluga, like most other sturgeons, is not a true freshwater species; it is anadromous, among the fish that are born in fresh water but spend most of their lives in salt water and then return to fresh water to spawn. For my quest, I wanted to focus on fish that lived their whole lives in rivers and lakes.

I had personal reasons for this. I had grown up around rivers and lakes, and while marine science interested me, when I thought of water, it was fresh water that came to mind. But it made sense from a conservation standpoint too. There was a greater urgency for the study of freshwater ecology than for any other habitat. About half of all fish species in the world are found in rivers, lakes, and wetlands, even though those areas make up less than one hundredth of one percent of Earth's water. As my experiences from the Mekong had shown, these ecosystems were degrading at an alarming rate, and there existed a critical lack of knowledge about what to do to protect them.

There was another reason: we already knew which species was the biggest fish in the ocean—the whale shark. For fresh water, it was still a mystery.

The catch of the 646-pound Mekong giant catfish in Thailand in 2005 generated a deluge of media interest. Seth Mydans of the *New York Times* wrote an article about it titled "Truly, It Was a Whopper, but Are There Bigger Fish?" Several other major media organizations—including the Associated Press, the *Los Angeles Times,* and *Science* magazine—picked up the story too. But the most important interest in terms of elevating the visibility of the Megafishes Project came from National Geographic.

A talented cameraman named Dean "Johno" Johnson had done a video story on me in Cambodia for the Associated Press. Johno was a partner in a start-up production company called InFocus Asia, based in Bangkok, that was looking to pitch TV ideas to National Geographic. Using footage of the Mekong giant catfish and the giant freshwater stingray, he cut together a short reel as part of a proposal for a one-hour documentary on giant fish with me as the host.

It was a good fit. The National Geographic Society had already supported my research in Cambodia, and in 2004 I had been named a National Geographic Emerging Explorer. I saw the documentary as an opportunity to share the stories of incredible fish that few people had ever heard about, and a chance for me as a scientist to make discoveries along with the audience.

National Geographic loved the idea.

Early on, we decided that the story should be built around the Mekong giant catfish. It was the story that I knew best. At the same time, there was a risk in picking this species. It had become difficult to find giant catfishes in the wild, and if we didn't find a large fish to film and talk about on camera, we wouldn't have a show.

After six months of planning, our first shoot was in Chiang Khong in northern Thailand, in the same area where the giant catfish had been caught in 2005. We made arrangements to stay in the last remaining fishing grounds for the giant catfish, in late April. We planned to stay until a fish was caught. But we also explored the area, talking with fishers, filming them and their traditions, which helped everyone become more comfortable with one another. That connection proved crucial, because when a fish was eventually caught, the fishers called us, and we were on the scene so quickly that the giant catfish hadn't even been removed from the net.

It was a good-sized fish—far from a record breaker, but a catfish larger than most people would have ever seen. As we filmed, I unwrapped it from the net and affixed a tag to it so that it could be tracked, before we pushed it out of the shallow water and released it into the deep. With a final push and a burst of speed from the fish, it was gone. It was the first time a wild Mekong giant catfish had been successfully tagged and released in Thailand.

Our luck continued as we traveled to Cambodia to get more footage, though this time we had to do some major improvisation. I had just arrived in Phnom Penh with an Australian cameraman named Wade Muller. After a late dinner, we returned to our hotel and, exhausted, went straight to sleep. At 1:00 a.m. my phone rang. It was my local research partner, Phanara. In his halting English, he excitedly informed me that a giant catfish had been caught at one of the dai fisheries. I looked outside. It was pitch black, and torrential rain lashed my window.

There was never any doubt that I would go. For Wade, it was different. He had not had time to organize his gear, or charge his camera batteries, and he had no lights to film in the dark. It seemed pointless to try to shoot under those conditions. So he stayed at the hotel while I left, alone, with my small camcorder and a flashlight. In the pounding rain, Phanara and I drove the project boat up to the dai fishery.

The fishers had caught a monster. It looked to be about five hundred pounds—one of the biggest fish that I'd ever seen. As we had done so many times before, Phanara and I immediately jumped into the muddy water with the fish and started to move it around to ascertain what shape it was in.

I gave my camcorder to one of the fishers to film us, while another man tried to illuminate the scene with my flashlight. By this time, the rain had subsided. I was focused on bringing life back into the fish, while explaining to the camera what we were doing. I gave little thought to how things would look on tape. I just made sure that the red dot on the camera was lit and that it was pointing in the right direction.

Normally, we would take the fish into the middle of the river to release it, but because the fish felt pretty strong, we decided to let it go closer to the riverbank. As I dove down with it and gave it a push, it swam away into deeper waters.

The footage turned out perfectly. Johno was ecstatic, and so was I. The whole program felt completely authentic. We had not staged a single scene, it was all based on the work that I normally did, and the science was accurate. I was happy to see Phanara and people from the fisheries department featured, and I know that many people in Thailand appreciated that the program had an American scientist who could speak their language explaining these rare creatures.

The response from National Geographic was equally enthusiastic. The documentary, which was called *Megafish,* aired to solid ratings. Immediately, there was talk about doing more shows, and maybe turning it into a series. The show was not going to be about finding the world's largest freshwater fish, but I knew it could aid me in my quest by providing financial resources and media exposure. I would be hosting my own National Geographic wildlife series, something I never dreamed would be possible.

The question was: what should the next show be about and where should we film it? The answer seemed obvious. The most natural place to look for large freshwater fish was in the world's largest river.

In describing the enormous size and abundance of the Amazon, one must put the numbers into context. The basin's square mileage, for example, is equal to the forty-eight contiguous US states. With an estimated 2.5 million different kinds of insects in the Amazon, a single bush may hold more species of ants than the entire British Isles. As for the river, the amount of water it sends into the ocean each day would be enough to supply New York City's water needs for nine years.

The magnitude of the Amazon rainforest came into clear view as we flew into Manaus, a Brazilian city of about two million people that sits in the heart of the basin. About 60 percent of the Amazon rainforest lies within Brazil, with the rest spread over nine countries. During the last couple of hours of my flight from Miami, I could see nothing but green jungle out of my window.

I had come to Brazil with a television crew—two producers, a cameraman, and a fixer—in search of giant fish. One of the giants we hoped to find was the arapaima (*Arapaima gigas*), a species that was not only on my megafish list but also had a good chance of being the biggest freshwater fish of them all. I had seen an old, detailed drawing of an Amazonian fisher sitting on top of an arapaima that appeared to be twelve feet long and of unusually heavy proportions, longer and just as bulky as the record-breaking Mekong giant catfish.

The arapaima is part of a family of fish called bonytongue, and it's found throughout the Amazon basin. In Brazil, the fish is called *pirarucu*. Legend has it that it was named after the disrespectful son of an Amazon chief who was struck by the god into the murky depths of the river to become the fish. What is definitely true is that *pirarucu* comes from the Tupí language and translates roughly as "red fish," a reference to its massive red-splashed tail. It has a scaly, torpedo-shaped body and a broad, bony head that is copperish green in color.

An extraordinary fact about the arapaima is that it breathes air and may survive up to twenty-four hours outside of water. It can stay submerged for ten to twenty minutes at a time, but then it needs to come to the surface to breathe. When it rises, it makes a distinctive coughing noise. With its small gills, it sucks in oxygen using a modified swim bladder that opens into the fish's mouth and acts as a lung.

However, the arapaima's need to surface to breathe makes it vulnerable to humans, who can easily target the fish with harpoons. Indigenous communities in the Amazon have long hunted the arapaima for food and for its large scales, which can be fashioned into jewelry and other items. Sought after for its flavorful meat, which can be preserved without ice, arapaima became known as the "bacon of the Amazon." As commercial fishing ramped up, arapaima populations plummeted, and by the turn of the millennium there was concern that the fish could disappear completely.

When I arrived in the Amazon, a nascent conservation effort had taken shape to turn things around for the arapaima. Brazilian and American researchers had begun working with indigenous fishing communities to establish new management practices and guidelines, and the initial signs were hopeful. It was early going, but it suggested that the protection of a giant fish like the arapaima was possible.

In Manaus, I got a chance to swim with the arapaima at a small aquarium outside the city. It's always a little unnerving getting into the water with a big fish for the first time. In my experience, every species reacts differently: some get skittish, while others don't seem to care that much.

The two arapaimas in the tank were big—at least as big as me—but they kept their distance. I could see their watchful eyes tracking my movements as I swam next to them, but I didn't feel threatened. I liked being in the water with them, watching their slow and graceful movements, appreciating their sturdy build, knowing they didn't react in a fearful or aggressive way. I got the sense that they viewed the tank as their home, and that I was a visitor they'd put up with until I got tired of holding my breath and would let them be.

As the commercial hub of the region, Manaus, which sits at the confluence of the dark (blackwater) Rio Negro and the pale, brownish (whitewater) Amazon River, is home to a massive fish market. Situated along a set of piers below the center of the city, the market was supplied by a constant flow of fishing boats that came to dump their catches after prowling the river system for weeks. It was a wild and chaotic scene, filled with traders shouting and gesticulating, and the diversity of fish on sale even exceeded what I was used to seeing in Southeast Asia.

As much as 30 percent of the world's biodiversity—that is, the total number of living things on our planet—is found in the Amazon. Out of roughly fifteen thousand freshwater fish species in the world, at least three thousand occur in the Amazon, and it is a number that is constantly rising. Scientists estimate there may be as many as fifteen hundred additional morphologically unique freshwater fish species that are yet to be discovered in the Amazon, and perhaps thousands more based on genetic differences.

Why is the Amazon so species rich? For sure, size has a lot to do with it. But location is also important. Ecologist Michael Rosenzweig, for example, has shown why larger geographic areas, particularly those near the equator, usually contain more species. Another key is temporal stability. Near-constant rain has made the system stable and predictable for a long time and helped guard it against volatile ecological changes, like prolonged droughts, that could have caused fish species to go extinct in the past.

Like the Mekong and Tonle Sap Rivers, the Amazon's aquatic ecology is governed by a productive flood pulse that each year submerges largely forest-covered wetlands. While the Amazon may look like a solid green carpet from above, up to one-sixth of the basin area is actually filled with water at certain times of the year. The flood pulse is the driving force governing all the ecological functions and interactions in the river basin, and the flooded forests that it creates are crucial for the survival and reproduction of fish.

In fact, there are few places in the world where aquatic and arboreal life are brought together as closely as they are in the Amazon. Hundreds of fish species feed on fruits and seeds floating on the surface of the water and on an abundance of plankton. Many fish also move into the flooded areas to seek shelter from predators, emerging from hiding once they have grown bigger. In turn, those forests, which have adapted to being under water for months at a time, also benefit from fish dispersing seeds throughout the system. The result is a mutually beneficial exchange of resources that fuels the Amazon's biodiversity.

Significant links between forest cover and fish abundance have been found for many of the Amazon's most popular food fishes, including the highly prized tambaqui, which can grow up to seventy pounds and has specialized, human-looking teeth that can crush and grind hard fruits and nuts. Such connections have also been established for many carnivorous species, many of them catfishes, which feed on schools of fish in the flooded forest.

It was in search of catfishes that we left Manaus on a seaplane bound for the Marmelos River, two hours to the southwest, to continue our filming.

During my years in the Mekong, I had developed an appreciation for catfishes. Many people, Americans especially, tended to lump all catfishes together as ugly and filthy bottom-dwellers, but that was not how I saw them. Catfishes range from adults an inch in length to species that can grow over ten feet long. They live in rivers and streams, in marshes, in bogs, in high-altitude lakes, in subterranean caverns, in hot springs, in the ocean, and under ice. They're equally varied in shape, color, behavior, and diet. There are wood-eating catfishes in Peru and electric catfishes in West and

Central Africa. For any type of water, any location, any temperature, and any food, there is a catfish to match.

In the Mekong, catfishes are the second-most diverse group of fish after carps. Yet their numbers there are nothing compared to the diversity that is found among catfishes in the Amazon, where some thirteen hundred catfish species have been identified. In other words, more than one out of every three fish in the Amazon is a catfish—including so-called goliath catfishes.

The biggest of these giants is the piraiba, sometimes called the lau lau. According to FishBase, the online database, the piraiba (*Brachyplatystoma filamentosum*) can grow almost twelve feet long and to more than four hundred pounds. It has large and spiky dorsal fins, and the fish's three pairs of barbels, of which the longest may reach almost the entire length of the body, give it away as a catfish.

Piraibas have a fearsome reputation as ravenous predators that will feed on most everything, including small monkeys, which have been found in the stomachs of fish that have been caught. But the contention that piraibas also attack and kill humans is most likely false. The claim stems in part from a dubious assertion made by US President Teddy Roosevelt during his expedition to the Amazon in 1912. Rumors, however, can be hard to kill. To this day, many local fishers still claim that people often go missing in the Amazonian waters where the piraibas are found, even though FishBase calls piraibas "harmless" to humans.

Instead, it is humans who present the bigger threat to the piraibas, which, like many other catfishes in the Amazon, are considered an excellent food fish. In the Brazilian-Colombian-Peruvian border region of the Amazon, the piraiba once represented the most important fishery by weight, with more than four thousand metric tons caught in 1988. By the time I got to the Amazon less than two decades later, the overall catch had plummeted to about one-eighth that level. Although the species had not been assessed for the IUCN Red List, it was clear it was being overexploited.

Despite the importance of Amazonian catfish to people, relatively little research had been done on them at the time. Numerous scientists were studying fish and freshwater issues in the Amazon—indeed that number was far higher than in the Mekong—but the Amazon was so vast and contained so many fish that studying all of them was a gargantuan task. Almost nothing had been done to manage the fisheries based on knowledge about catfishes, for the simple fact that so little was known about their life cycles.

There was, however, one groundbreaking study going back to the 1970s that had yielded some fantastic findings about the migratory patterns of the

Amazonian catfishes, in particular the dorado (or *dourada,* as it's known in Brazil), a cool-looking, gold-colored species with a platinum head that can grow longer than six feet, which has also been heavily fished in the Amazon.

At the time, there was no solid evidence that fish—or any other animals, for that matter—required the expanse of the Amazon basin to meet their reproductive or other ecological needs. In the case of the catfishes, no one had seen or reliably reported the whereabouts of spawning grounds or larval fish. So, in the 1970s, ecologists Michael Goulding and Ronaldo Barthem set out to find out.

Because the Amazon was too large for tagging experiments, Goulding and Barthem conducted their research the old-fashioned way: by identifying and measuring fish in markets, on commercial fishing boats, and during scientific expeditions in different parts of the river basin.

After almost two decades spent analyzing sixty thousand specimens and the stomach contents of more than ten thousand fish, it became clear that the catfishes indeed traveled enormous distances within the vast river system, and that spawning took place mostly, if not exclusively, in the western regions of the Amazon. The researchers also found evidence for the dorado's elusive spawning grounds—in the form of larval fish—higher up the river system than anyone had thought, in the foothills of the Andes mountains, an astounding thirty-six hundred miles from the mouth of the river.

It meant the dorado travels more than seven thousand miles to complete its life cycle, starting upstream and drifting all the way to the floodplains and estuary, where it spends most of its adult life before eventually moving all the way back upstream to spawn. This epic journey is by far the longest known migration of any freshwater fish species in the world.

The long-distance migratory behavior of the catfishes, Goulding and Barthem argued, was probably the main reason for their historical abundance. But, as they pointed out, the same migrations also made the catfishes more vulnerable to humans.

The show had been retitled *Hooked! Monster Fish,* and now we needed to hook some monsters to film. The plan was to join a mobile fishing camp for recreational anglers on the Marmelos River and use it as a base from which to go fishing for goliath catfish. But the fishing camp was fully booked with clients, which not only meant our crew had to sleep in tents on floating pontoons but also that there were no fishing guides available. We would have to do the fishing on our own.

The Marmelos River, or Rio dos Marmelos in Portuguese, is a tributary of the Madeira River, which in turn is the largest tributary of the Amazon River. With its headwaters in a national park that had been established in 2006, the Marmelos runs through rainforest, grassland, and savanna. Sparsely populated by humans, it is an area renowned for its diverse fish fauna. A survey by Brazilian researchers a few years prior to our visit recorded 133 fish species inside the Marmelos Conservation Area, with four species endemic to that very place, meaning they're not found anywhere else in the world. I had not seen the survey, unfortunately, so I didn't know that the researchers had not recorded any goliath catfish.

It quickly became clear that we wouldn't find any goliath catfish while fishing at the mobile camp, especially since we didn't have any local guides to help us. Fighting swarms of biting flies and mosquitoes, rod-and-reel fishing produced a few peacock bass. But for the most part I kept hooking piranhas.

There are few fish with a more ferocious reputation than piranhas. The image of piranhas moving in shoals and tearing to shreds any flesh that dares to dip into the waters is seared into many people's minds. The piranhas' fearsome image had been stoked by none other than Teddy Roosevelt, who wrote in the memoirs of his Amazon trip how piranhas "habitually attack things much larger than themselves" and "devour alive any wounded man or beast." He recounted watching a pack of piranhas devouring an entire cow.

In fact, piranhas are hardly the vicious freshwater monsters they are made out to be. They do have very sharp teeth and fierce bites—the word *piranha* translates to "tooth fish" in the Brazilian language Tupí—and many are carnivorous. But among the sixty or so species, some are omnivorous and eat more seeds than meat. There's little evidence to support their reputation for attacking people or large animals, unless those people or animals are already dead or injured. Even President Roosevelt's story has a much more mundane explanation: It appears that the locals had put on a show for old Teddy by tossing an already dead cow into the river, and then releasing the pack of piranhas, which they had caught and kept in a tank without food.

The fish that I kept hooking were red-bellied piranhas, which are the second-largest species of piranha and grow about a foot long. Considered the most dangerous and aggressive of the piranhas, they have an interesting feeding technique. The fish spread out and a "scout" signals when a food source is found. When feeding, they are very orderly, with some of the fish taking a bite before moving aside so that others can get a portion.

It was easy to tell when piranhas had found the bait. Violent knocks on the end of the line—*whack! whack! whack!*—meant piranhas in the school were taking turns biting off chunks of bait with their powerful jaws. If I was using a large hook, large bait, and wire fishing line, they'd simply take chunks off the bait until there was nothing left, leaving me to reel in a bare hook. If I used smaller hooks and bait, I would hook the piranhas one after another. They fought aggressively when hooked, moving this way and that in quick pulls on the line. If they were well hooked, I could bring them in to the boat in a few seconds, but getting them off the hook was tricky. They could easily twist and bite my finger, so I was always careful and used pliers rather than my bare hands to dislodge the hook.

A couple of days had passed when we were told about a remote fishing spot in an upstream tributary, near a waterfall, where people had been catching jau, or gilded catfish, a widely distributed, bottom-dwelling species that may grow up to two hundred pounds and that likes to hang out in deep holes and fast currents. I had heard about the jau and its habit of attacking various migrating characins (the large group of fish to which piranhas and others belong) as they attempt to negotiate rapids. The jau had evolved to handle living in rough waters without being battered to death.

We commissioned a motorized canoe and set off for the waterfall in the morning on what was supposed to be a day trip. The place turned out to be a lot farther away than we thought, however, and when we finally arrived at the waterfall it was already afternoon. We had no choice but to stay overnight. We had brought no food or even tents to sleep in, so would have to catch our own dinner and sleep outside on a sandy beach next to a large rock formation. It could have been worse. The spot was beautiful, surrounded by pristine rainforest, and we would have no trouble catching piranhas to eat that evening.

But how were we going to catch jau? Or any other fish that wasn't a piranha. That's when I thought of a trick I had used on a fishing trip years before, when I had left a line with a baited hook in the water overnight. Maybe that was a method that could work here? The piranhas appeared to spend most of their time high up in the water, in a different space from the catfishes. They fed on sight rather than smell and seemed to be less active at night. I thought that if I got a strong line and a heavy bait to put on a hook, making sure it stayed at the bottom, it could work to lure a catfish without attracting the piranhas swimming above.

It wasn't the most sophisticated of fishing techniques, but I've always been of the opinion that you use whatever works, whether you're fishing or doing research. So I attached the line to a pole and stuck it in the sand. On top of the pole, I placed a bell, which would wake us up to alert us of a catch, and then I went to sleep.

I'm not sure how long I had been sleeping when I woke up from the bell ringing. I jumped up and ran to the pole and pulled on the line. There was definitely a big fish on the other end, but not a jau. Instead, it turned out to be a surubim, or striped catfish, another big fish that, like the jau, prefers to stay at maximum depth in the river. It was a beautiful specimen, almost four feet long with long whiskers and a striking pattern of spots and stripes along its body. Not a goliath, but still a special catch, I thought, as I released it back into the river.

It was impossible not to be swept up by the vastness and the unspoiled beauty of the Amazon. On a river like the Marmelos, it was as if time had stood still. It flowed as it always had through a jungle lush and impenetrable. Or so it seemed.

In reality, things were changing in the Amazon. From our vantage point, the forest looked intact. But beyond its riparian band, it was disappearing. Up to twenty-seven thousand square miles of Amazonian rainforest— almost the size of the state of Maine—were being lost in Brazil in just one year, to cattle ranchers and soya farms. On the Madeira River, illegal gold mining operations were proliferating, dredging and destroying the river structure forever. And while the Amazon remained largely undammed, there were plans for hundreds of hydropower plants to be built.

There were plenty of fish in the river system, even if we had had trouble catching goliaths, and fisheries played an increasingly important role not only in providing protein to the Amazonian population but also in supplying urban demand. People were eating more fish, and fish stocks were going down, especially for migratory fish. Production was still almost four hundred thousand tons per year, but for how long could such a yield be sustained? To me, the Amazon system looked pristine. But what was I comparing it to?

At the time, "shifting baselines" was a relatively new concept that had taken hold in a variety of scientific disciplines. The term had been coined by a fisheries biologist, Daniel Pauly, and it could be described as the phenomenon where each successive generation assumes that any given diminished biological state is the norm. Over time, knowledge is lost about the state of the natural world, because people don't perceive the changes that are actually taking place. It was a simple but very powerful concept that explains our generational blindness to environmental destruction and biodiversity.

Pauly had developed the concept in reference to marine fisheries science, where the problem was plain to see. Fisheries scientists often accepted as a baseline the stock size and species composition existing at the beginning of their careers, failing to take into account how abundant a fish species population had been before human exploitation, or what could be called the "true" baseline.

But the situation was even more complicated for freshwater fisheries, which often lacked basic information about species populations, present or past. This was in large part because inland fishing is made up of a lot of artisanal, recreational, and illegal fisheries, and it is difficult to use catch statistics to assess stocks. In fact, according to the UN Food and Agricultural Organization, catches from inland fisheries in 1999 were underreported by a factor of three or four, and there was no way of telling what that discrepancy might have been going farther back in time.

In the Mekong, I had seen giant fish species becoming rarer and increasingly forgotten. Many people didn't even know they existed. This was largely due to a lack of documentation of past abundance. If people didn't know how plentiful species like the Mekong giant catfish and giant *Pangasius* had been in the past, was it any wonder they didn't know these fish were quietly drifting toward extinction?

While the overexploitation of the world's fisheries had become the subject of much concern by the turn of the millennium, the discussion had focused nearly exclusively on marine resources and related charismatic animals such as seabirds, marine turtles, dolphins, and whales. Fisheries in inland waters, on the other hand, had received only slight consideration within global analyses.

It was a point made in a paper on the overfishing of inland waters that was published in the journal *BioScience* in 2005, and for which I was a coauthor. The lead author, a well-known freshwater ecologist named David Allan, had contacted me at the Center for Limnology at the University of Wisconsin, Madison, where I was working, to see if I could write the paper's sections about the Mekong River and the Tonle Sap Lake. They were perfect examples of inland overfishing. Fishers on the Tonle Sap had reported that catches of large catfish had dropped by 90 percent—or in some cases even 99 percent—in some fishing lots.

But the overexploitation of fisheries was not always marked by declines in total yield. One of the symptoms of intense fishing in inland waters was the collapse of particular stocks even as overall fish production rose. In other words, it was a biodiversity crisis more than a fisheries crisis. "This should be of broad concern," we wrote, "yet threats to freshwater fisheries and associated biodiversity have received scant attention from conservation groups and the media. This imbalance seems particularly dangerous considering evidence that freshwater ecosystems...are, on average, more threatened than marine ecosystems."

There were so many more people working on marine issues than on freshwater matters. For every peer-reviewed article that focused on fresh water, there were twenty studies on marine issues. The point was brought home when, in 2005, I was invited to the UN Convention on Migratory Species (CMS) in Nairobi, Kenya. Among the many fish experts gathered there, almost all worked in the marine realm, and there were a lot of discussions about marine turtles, seabirds, and sharks.

I had just completed my dissertation work on the Mekong giant catfish, which I presented at the meeting. To my surprise, I was asked if I would be willing to serve as "Councilor for Freshwater Fish," an unusual appointment for someone so early in his career and an indication that there were few experts on global trends regarding freshwater fish. I realized that it wasn't just the public who had not been told the megafish story.

I had swum with arapaimas in the aquarium in Manaus. Now I wanted to see them in the wild. For that, our crew headed to the mouth of the Amazon River, as it empties into the Atlantic Ocean. While arapaimas are not migratory fish, their movements and reproductive cycle are highly dependent on the Amazon's seasonal floods, and they are found all over the river basin, including in its estuary.

When the rivers of the Amazon overflow, arapaimas are dispersed into floodplains containing so much decaying vegetation that oxygen levels are too low to support most fish. Then, during low-water months, arapaimas construct nests in sandy bottoms where the females lay eggs. Adult males play an unusual reproductive role by incubating tens of thousands of eggs in their mouths, guarding them aggressively and moving them when necessary. The eggs begin to hatch when water levels rise, providing the conditions necessary to flourish.

We stayed in a resort on Marajó Island, which sits next to Mexiana Island, the largest island at the mouth of the river (and which I was told is as large as Switzerland). Even Marajó didn't feel like an island, more like an endless floodplain of submerged canals and streams and dotted with palm trees.

As it turns out, the resort, which had mainly housed recreational anglers, was on the brink of closing. The owners had decided to get into the arapaima aquaculture business instead. Farming arapaima was becoming increasingly popular in Brazil, with demand for its meat steadily rising among urban eaters and from abroad. The industry was still in its infancy, and there was still a lot to learn about the fish in order to support and expand production. But the arapaima grew fast and big, which made it a very good candidate for farming.

In fact, the arapaima's growth rate is the fastest among known freshwater fish species, with juveniles gaining more than thirty pounds a year after birth. Its food conversion rate is the highest recorded in any fish, with young arapaima retaining as weight as much as 70 percent of all food they consume. In 2018, the arapaima became one of the first freshwater fishes to have its genome sequenced, with the study yielding important clues about gigantism and fast growth in fish. In their genetic analysis, the researchers showed that the functions of more than a hundred of the arapaima's genes are associated mainly with muscle development, contributing to its enormous body size.

On our first day in Marajó, we visited the aquaculture operation the hotel owners had started. The operation appeared fairly low-tech: a series of earthen ponds and canals that did not look very different from the surrounding, swampy landscape of water hyacinth and other weedy aquatic vegetation. The owners used at least one building for captive breeding, and they showed us young arapaima of various sizes that were kept there. Bred in captivity, the young were raised to a certain size, after which they would be released in the ponds to grow and eventually be harvested.

Back at the resort that evening, we had dinner in a palapa, a tall, open-sided hut with a thatched roof made of dried palm leaves, a useful construction in hot weather. There were no other diners, and all that was on the menu was arapaima in various forms—grilled or breaded, turned into nuggets, or put on fish sticks. I don't normally eat the fish I study, but this time I made an exception, since I knew the fish had been raised sustainably. It was tasty.

On the Marmelos River, our main problems had been that we didn't have someone local to guide our fishing efforts. In Marajó, the lodge put us in touch with a local fisher named Fernando, an older gentleman who had been fishing for arapaima for many years and promised to help us catch them.

Fernando took us along a river channel that cut through grassland and led to a pond. Arapaimas had evolved in the world's largest river system, but they could mostly be found in smaller, confined areas. Once we arrived at the pond, we cut the boat motor and sat silently, listening for the telltale sound of arapaima coming to the surface to breathe, which is when traditional fishers catch them with harpoons. We did not use harpoons but a rod and reel, equipped with heavy-duty braided line and large, fish-shaped lures.

The arapaima seemed skittish, so we tried to be as quiet as possible, avoiding any movement or unnecessary noise. We cast to each area of the pond rather than making a disturbance moving the boat. Once hooked, arapaima react with an initial burst of energy, but tire more quickly than a similarly sized catfish. Even so, a big arapaima can require a lot of strength and time to bring to the boat.

We had been fishing for a while when someone hooked an arapaima after casting in the spot where it had surfaced. Judging from the bend in the rod, it was big but not a record breaker. When we got it to the side of the boat, I jumped in the shallow water to get a closer look. This can be dangerous because arapaima have hard, bony heads and they can thrash around or jump in the water, swinging their heads and injuring fishers.

Luckily, I was able to handle it without incident. The arapaima was about four feet in length. Fernando reckoned it was only three or four years old. But since the arapaima is such a fast-growing species, it wouldn't be long before it turned into a giant. It had already grown to fifty-four pounds. After taking the measurements and filming it, we released it back into the water, careful to give it time to recover from capture. Releasing an air-breathing arapaima too early can be dangerous. An exhausted fish may sink to the bottom and, lacking the energy to come to the surface, drown.

It was a nice catch, but naturally we wanted to catch a bigger fish. After all, the show was called *Hooked: Monster Fish,* so we had to hook a monster. But the fishing wasn't just for the show. The point was to learn firsthand about these giants, and to do that, we had to catch them first. Fernando offered to show me his local fishing grounds and his technique.

Fernando didn't use a rod and reel to fish for arapaima, but rather a trotline, a heavy fishing line with smaller lines and baited hooks attached at intervals. With as many as a hundred hooks per trotline, up to a hundred fish can be caught at one time, but typically the catch is a lot less than that. Tying the end of the line to a tree on one side and to a heavy rock on the other, Fernando fished the entire width of the river, approximately one hundred meters.

Given the amazing diversity of fish in the Amazon, it was impossible to predict exactly what we would catch. But as soon as we began to pull up the line and check each hook, I could feel the weight and tug on the rope, indicating there was something big hooked farther down the trotline. As we pulled in more rope, it became heavier and heavier, until a fish came to the surface. At first glance, I couldn't tell what it was, only that it was as wide as it was long, with a tail that seemed to be twisting and dancing around the bait line. Fernando quickly pulled it into the boat, and that's when I realized we had caught a freshwater stingray.

Before I could get a good look at this fish, with a practiced movement Fernando whacked it with his machete. In an instant, he chopped off its tail, which held its stinger, and threw it overboard. It was shocking to see, and seemed unnecessary. But he later explained he had been injured badly by mishandling stingrays, and he didn't want to take any chances, especially with a guest on board.

We caught other fish, the most significant being a small piraiba, the catfish that Teddy Roosevelt had suggested could kill a human. The piraiba we caught was only about three feet long, so probably a juvenile. With its sharklike appearance, it was still a cool-looking fish, and considering its decline in the Amazon, it was encouraging to find it. One day it might grow to four hundred pounds, which FishBase reported as its maximum size. I hoped so, as we released it back into the water.

It was late in the day, and I was starting to doubt that I would encounter another arapaima. We had checked all the trotlines, except one. As we pulled on it, I could feel heavy resistance. As we continued to pull, it got harder and harder, until suddenly I spotted the characteristic green and red colors of the arapaima's scales in the muddy water. It looked big, and even Fernando seemed interested. He quickly helped me pull it into the boat.

It was an arapaima about six feet in length. It wasn't close to full size, but it was big enough to be harvested. As I looked at Fernando, I could detect a moment of hesitancy. Did he want to kill it? He shook his head. No, let's film it and release it, he said. After handling it for a few minutes, we slipped it gently back into the waters. It was a gracious gesture by Fernando, and the fact that he had helped us catch several of these fish, including a fairly large arapaima, underlined the importance of working with people who had local knowledge about the rivers. The biggest fish we caught in the Amazon was thanks to someone who had fished there for many years.

In the end, it was a successful trip. We had caught three of South America's largest fish species in one day, and National Geographic was happy with the resulting program. Even so, I left the Amazon with more questions than answers. The arapaimas that I had seen were not as big as the ones I had heard of. Maybe I was just being naive to think that I could find a giant on a two-week trip. Or perhaps it was something else. Maybe the arapaima, through years of being hunted for its meat, had not only seen its numbers diminish but also been cut down in size?

Still, I felt like I was on the right track. I may have been no closer to answering the question of what was the world's largest freshwater fish, but asking it was what had pushed me out into the world in search of answers. The Amazon had made me realize more than before the scope and scale of my quest. Maybe there were some things I had to rethink about the project. For one, the idea that the size of the system, its diversity and productivity, would yield the biggest fish was not necessarily true. And while big rivers are home to big fish, I realized that I shouldn't limit myself to the largest sections of the largest rivers. Very large fish might be found anywhere, even in places where I wouldn't expect them.

Maybe it was not just a question of where to look, but also when. For some time, I had been wondering where these giant fish came from. As I started to learn more about them, a troubling notion haunted me. Many of the giants that were in such trouble were ancient fish that had been around for hundreds of millions of years, and now they were fading out of existence on our watch.

Old Fish, Ancient Origin

To understand how giant fish came to live in our rivers and lakes, we must travel back in time, more than 540 million years, to the most important evolutionary event in Earth's history. Picture a relatively tranquil planet covered in shallow, oxygen-starved seas with slimy mats of microbes. On those mats existed an assortment of blind and mostly stationary animals with bodies that looked like thin pillows. Without predators, it was a safe world for everyone, if a bit dull.

Over the next ten million years, however, a small increase in oxygen led to a burst of evolution called the Cambrian explosion. New and complex creatures with bones, shells, and organs with specialized functions, like eyes, gave rise to carnivory, which in turn ramped up an evolutionary arms race that continues today.

Animals with backbones—chordates, a group to which not only all fish, but also reptiles, birds, amphibians, and mammals today belong—were later joined by ostracoderms, a collection of small vertebrates no bigger than six inches that were covered with external skeletons and looked like tadpoles with stone helmets. Then, one hundred million years later, in the wake of the first of five major extinction events, a new group of vertebrates— placoderms—appeared with a fancy new anatomical feature: the jaw.

Quite possibly the most significant innovation in vertebrate history, the jaw allowed animals to engage in both predation and defense. Placoderms, therefore, played a significant role in fish evolution, and some actually looked like fish, complete with several sets of fins. Yet placoderms are over-shadowed by the later arrival of two families of jawed fishes, more than four hundred million years ago: osteichthyans, or bony fish, and chondrich-thyans, cartilaginous fish that include sharks. All fish today can trace their ancestry back to these two groups, which emerged during the Devonian period, or the so-called age of the fishes.

At this time, true freshwater environments appeared. Until then, much of the planet's surface had been covered by epicontinental seas, with little solid land formation. But then came vascular plants, which, unlike previous aquatic plants, developed stems, leaves, and roots on land. This vegetational expansion solidified the land and created a totally new biome with streams, rivers, and wetlands that linked the terrestrial environment more closely with the aquatic realm and created a stronger distinction between freshwater and marine habitats.

Before then, fish had evolved mostly in the confines of coastal saltwater lagoons, with large parts of the seas devoid of life. Now, they could move into dynamic new habitats that offered protection as well as feeding and breeding space. As many fish underwent evolutionary adaptations for their new lives in fresh water, bony fish split into two groups: lobe-finned fish and ray-finned fish. The former had fleshy fins joined to the body by a single bone, whereas ray-finned fish are so called because their fins are webs of skin supported by bony or horny spines.

The early freshwater worlds were an evolutionary cradle for fish, which helps explain why about half of all fish species today are found in fresh water. All of the early sharks, for example, evolved in fresh water. So did the majority of the ray-finned fishes, which today make up more than 90 percent of fish species.

For tens of millions of years, the aquatic fauna kept building up, with all the different fish groups expanding. But eventually life got crowded and competitive, pushing some fish to the land. Around 367 million years ago, lobe-finned fishes such as the four-foot-long *Panderichthys* and the similarly sized *Eusthenopteron* had evolved into the first tetrapods, the proverbial "fish out of water" ancestor to all land-living vertebrates, including humans.

After the second of five major extinction events, a hundred-thousand-year-long cold spell triggered the growth of glaciers almost down to the tropics, making life difficult for fish. With sea levels falling substantially, huge ecosystems were wiped out, including an estimated 85 percent of aquatic species. For the next eighty-five million years, massive tectonic activity and the formation of the supercontinent Pangaea resulted in the creation of mountains and new rivers and lakes—and the evolution and expiration of many fish species.

But 250 million years ago, it all (almost) came to an end during the end-Permian extinction—also known as the Great Dying. During Earth's most severe known extinction event, caused by a series of volcanic eruptions in Siberia that caused huge amounts of carbon dioxide to be released into the atmosphere, up to 96 percent of all marine species and 70 percent of vertebrate species on land were destroyed. But evidence suggests that freshwater fish fared much better than their marine counterparts. Deposits discovered in present-day South Africa show freshwater species coming through the end-Permian practically unscathed, lending credence to the idea of freshwater systems as a kind of refuge for fish.

The Triassic period that followed the Great Dying was a time of tremendous change and rejuvenation. Disaster fish like *Saurichthys,* or "lizard fish," dominated the waters at first, but soon a slew of new reptiles and amphibians slunk and slithered on and off the Triassic coast, lakes, and rivers. Then, 230 million years ago, the first mammals—tiny shrew-size creatures known as morganucodontids—appeared, and not long after that came the first dinosaurs.

When a fourth extinction event occurred about 200 million years ago—caused perhaps by a volcanic belch or an asteroid collision—the dinosaurs survived and came to dominate the Jurassic era that followed. As tectonic forces broke up Pangaea, much of the evolutionary race was playing out on land. But there was a lot going on beneath the surface too. In fact, it was during this time that the ancestors of many of today's river giants first appeared, among them a group of predatory fish with thick, bony skulls and heavy scales known as gars.

One hundred fifty-seven million years later, I found myself in the Louisiana bayous in search of one of their descendants.

I had been told that if you wanted to catch alligator gars in south-central Louisiana, the guy to see was Ricky Verrett. A man in his early forties, he had grown up south of Houma, a city to the southwest of New Orleans that is named after Verrett's Native American tribe. In 2009 I met him there, on his fishing boat anchored in one of the many channels that make up the Louisiana delta.

In my work I've encountered innumerable fishers, but few as colorful as Verrett, a stocky man with a thick Cajun accent and a magnificent mullet. Like many other Houma people, he had devoted himself to a life on water. Verrett didn't own a car, and didn't bother much with things like shoes. He wasn't married because, as he put it, women were too much trouble, and besides, he liked fishing too much. Twice a year, he took his trawler out to fish for shrimp during south Louisiana's shrimp seasons. He also kept a crawfish pond. But it was the hardy alligator gar that provided his main income.

The alligator gar, which is the largest of seven existing gar species and can grow to over three hundred pounds and more than ten feet long, may be North America's largest freshwater fish. Since an early age Verrett had been catching these green-and-silver torpedo-shaped monsters, as one of the few people who fish alligator gar for food. It's not that they're not good to eat; alligator gars have plenty of good-tasting meat situated along their backbones. The challenge is cleaning them to get to that meat. The gar's thick, overlapping scales, known as ganoid scales, can't be scraped off; they have to be cut off with tin snips. Most people can't do it, but Verrett claimed he could skin a gar in twenty seconds flat.

He agreed to show me how he caught them. Accompanied by Dr. Allyse Ferrara, a biology professor at Nicholls State University in Thibodaux who studies the alligator gar and introduced me to Verrett, we set off in his open motorboat and headed into the bayou. Many people think of the bayou as thick, almost impenetrable swamp forest, with Spanish moss hanging from cypress trees, and there are areas that look like that. However, farther south, it feels more like an estuary, with braided channels cutting through treeless grasslands. It reminded me of Tonle Sap Lake in Cambodia.

The water here, closer to the sea, is murky and brackish. Although the alligator gar is considered a freshwater fish, it can tolerate a fair amount of salt in the water, up to seven or eight grams per liter, according to Ferrara. During his shrimp runs, Verrett had even seen alligator gars around the Louisiana barrier islands.

It was long believed the first gars appeared in freshwater environments about one hundred million years ago. But the discovery of cranial remains in Mexico has extended the record of Lepisosteidae, the family to which gars belong, back by an additional fifty-seven million years. Those gars evolved from a class of fish known as holosteans, meaning "whole bone," in reference to their notoriously bony skeletons, and at least ten species of these prehistoric gars have been identified. The fifth mass extinction that occurred at the end of the Cretaceous period, sixty-five million years ago, which wiped the dinosaurs off the face of the Earth, didn't destroy the gars. In fact, they spread to every continent except Australia and Antarctica.

Of the seven species of gars that exist today, the alligator gar was the first one to emerge, about five million years ago, and it's bigger than all of the gars of the very ancient past. It can be distinguished from other gars by the two rows of teeth in its upper jaw, versus the one row found in the others, as well as its wide head, which is similar in shape to an alligator's when viewed from above.

But gars have actually changed little in both appearance and physiology since they first emerged, and there are obvious reasons why. Like many living ancient species derisively referred to as "primitive," hidden within their prehistoric bodies is a string of evolutionary adaptations that have served these river giants well for millions of years. They are opportunistic feeders who primarily prey on fish, but also eat crabs, turtles, birds, and small mammals, and their spiral guts can wring the maximum nutrition out of kills. The gar's eggs are toxic, serving as a defense mechanism against potential predators. (The eggs are also poisonous to humans if ingested, another reason why gars are not popular food fish.)

Like the arapaima, gars can breathe air from the surface. In fact, the fish may obtain as much as 70 percent of the oxygen it needs from the atmosphere; their spongy and highly vascular air bladder behaves like a lung to aerate the fish's blood. This lets them live in places where most other fish can't survive: warm, sluggish waters low in oxygen, like those in the Louisiana bayou.

Yet today gars are found only in North America, and while the range of the alligator gar once included the entire Mississippi River basin, as far north as Iowa, they are now known to live only in the southern belt of the United States, from Texas and parts of Mexico in the west, to Florida in the east, with an isolated population in Cuba. Not surprisingly, humans have everything to do with that decline.

Alligator gars are venerated among several Native American tribes, including the Houma and the Coushatta, with the fish even depicted in the latter's tribal seal. There is a long tradition of Native Americans using gar scales to make jewelry and tools, leather products, and skin oil for insect repellent. Verrett himself has a giant pile of arrowhead-shaped scales, and his cousin sells traditional crafts made from gar scales at festivals.

But the alligator gar's reputation among anglers has historically been far less stellar. Throughout the twentieth century, gars were dismissed as "trash fish" that damaged nets and devoured game fish like bass and perch. Resource managers recommended culling them, and fishermen shot them with bows, clubbed them with paddles, and threw their carcasses up on the banks to rot. In the 1930s, officials even used "electric gar destroyers" to systematically exterminate the fish, and in several states it was long illegal to return gars to the water alive once they'd been caught.

Those removal efforts, combined with habitat loss, pushed the alligator gar deeper into the swampy rivers and brackish bayous of the South. At the end of the twentieth century, only Texas and Louisiana had stable populations, and even there the gar had no legal protections, with people free to catch as many as they liked.

Verrett obviously did not consider the alligator gar a "trash fish." "Not to me they're not," he said as we emerged from a channel into a shallow marsh lake the size of a dozen football fields. We had been traveling for a while through a maze of waterways that to my untrained eye looked easy to get lost in but that Verrett clearly knew like the back of his calloused hands.

The fishing method he employed, and which his dad had taught him, was not exactly hi-tech. To catch the alligator gars, he used hooks baited with fish and tied to strings of twine, which in turn were attached to empty milk jugs brightly painted orange that he would leave floating on the surface of the brownish water. If the milk jugs started moving, then a fish had been hooked.

Verrett brought twenty such contraptions with him in the boat. Usually he set more in the late afternoon and returned the next morning to collect his catch, but this day we were going to do it all in one go. Pulling close to the edge of the lake, he threw in the first jug, which quickly steadied on the surface. I was surprised that the jugs weren't attached to anything; it seemed like a hooked fish could just swim away with the jug and not be found again. But Verrett explained that the gars more or less stuck to a territory, and that swimming far away while pulling the jugs under water would quickly tire out the fish.

We set four or five jugs around the lake before moving through another channel and into a second lake, where Verrett did the same thing. At least an hour, but probably two, passed before all the jugs had been placed around several small lakes, and we headed back to the first lake. From a hundred yards away, Verrett immediately spotted one of the orange floats bobbing up and down and being towed across the surface. "There we go," said Verrett with a grin, and steered the boat in the direction of the moving jug.

As he caught up with it, he stopped the boat and lurched over the gunwale to grab a hold of the jug. Immediately, there was resistance on the other end. Verrett pulled on the twine, slowly inching his way down until eventually a thick, flattened head breached the water's surface. It was an alligator gar all right.

There has never been a verified gar attack on a human, but you have to be careful when pulling them into a boat where they can whip their bodies around and knock you off your feet. Equally hazardous are their teeth. With a mouthful of needle-sharp chompers, the alligator gar is among the most ferocious-looking fish around. Getting a hook out of its mouth requires extreme caution, because a bite can cause real damage. Verrett took no chances. Holding on to the line with one hand, he hooked the claws of a battered hammer into the fold of skin under the fish's jaw and hauled it aboard, quickly delivering a flurry of blows to its skull.

It was not a particularly large specimen, maybe four feet long. Looking at it, it struck me just how odd of a fish it was, with its thick and cylindrical body. It looked like an ancient creature, all teeth and bone and scales. Yet despite the gar's ability to persist through mass extinctions and outlive the dinosaurs, this fish was now staring up at me from the bottom of the boat as if to ask, "I survived all that for this?"

Verrett caught another half-dozen gars that day. They all turned out to be about the same size, no more than fifty pounds or so. Although few people fished for the gar in Louisiana, it seemed probable to me that harvesting had taken a toll on the size of the alligator gar, just as it had done with the arapaima.

I always had the idea that fish were bigger in the past. As a kid, I was mesmerized by illustrations in books about dinosaurs and the Jurassic era, and the aquatic creatures that I remember from those illustrations were always gargantuan beasts ready to leap out of the water to swallow a pterosaur whole. Many of the river giants that I studied as an adult, like the gars, had ancient lineages. Was it a coincidence that many of the very largest freshwater fishes were also very old?

We know that everything on Earth started small. We also know that organisms in evolving lineages tend to increase in size. This phenomenon is known as Cope's rule, after the American paleontologist Edward Drinker Cope, who in the late 1800s showed that the body sizes of terrestrial mammals such as horses generally increased over evolutionary time. Although there are exceptions to Cope's rule, like birds and insects, the fact that the blue whale, the world's largest animal today, is also the largest to have ever lived on Earth—bigger than all the dinosaurs—seems to shore up Cope's rule.

But Cope's rule does not mean that there weren't bigger fish in the past, since not all evolutionary lineages survive. In fact, more than 99 percent of all species that have ever lived on Earth have gone extinct, the majority of them during mass extinction events. Survivors, like the gars and other ancient fish, are the exception to the rule.

How many giant fish species have evolved and died out in the ancient past? To answer that question, we turn to the fossil record. But often fish paleontologists must work with fragmentary fossils, as in the case of *Leedsichthys problematicus,* a ray-finned, plankton-eating giant that lived in the oceans of the Middle Jurassic period, and was long believed to be the largest fish that ever lived. It was first described in 1889 from partial fossil remains. Almost one hundred years later, in 1986, paleontologist David Martill analyzed those remains and, citing the fish's huge "gill basket," concluded that *Leedsichthys* might have been at least ninety feet long. He dubbed it "the world's largest fish."

But was it? A few years ago, scientists reexamined that claim, turning to a wider collection of *Leedsichthys* remains, including a partial skeleton that had been recently excavated, as well as skeletons of closely related fish. The new analysis showed that the species may in fact have grown about only half as big as the previous estimate. It was huge, but probably not the largest fish that ever lived.

Instead, that title may belong to a shark called the megalodon, or the "Meg," as the 2018 Jason Statham movie would have it. This awesome predator—scientific name *Carcharocles megalodon,* not to be confused with a genus of clams also known as *Megalodon*—had teeth as large as a human head and could grow more than fifty feet long, roughly three times the length of a modern-day great white shark. The megalodon has been well studied, in large part because of its gigantic size and fearsome reputation, but also because it lived in relatively recent times, appearing twenty-three million years ago before mysteriously disappearing twenty million years later.

The study of other ancient sharks is complicated. Not only does the fossil record of sharks go back very far—420 million years—it is also problematic to establish because shark cartilage doesn't easily fossilize. Could there have been bigger sharks before the megalodon? Maybe. Since sharks evolved in fresh water, it's even possible that the largest shark—or the largest fish, period—that ever lived was a freshwater shark. Unlikely perhaps, but not impossible.

But if we stick to what we know today, what is the largest freshwater fish that ever lived? For that, we go back to when the Devonian period, the "age of the fishes," was ending. Life below the surface, especially in freshwater environments, had not only grown crowded but also big, and it included a group of large-bodied, predatory, lobe-finned fish called rhizodonts.

Large aquatic fauna did not do well in the wake of the Devonian extinction, and many disappeared as small, fast-breeding ray-finned fishes took over. But there were a few large-bodied, slow-breeding survivors, among them the rhizodonts. From that group one particularly enormous fish species appears to have emerged in the early Carboniferous era; fossils of it were discovered during the "fossil gold rush" that took place in Scotland and the north of England during the 1830s.

This species, *Rhizodus hibberti,* lived in fresh water, and it was a monster. Exactly how large it grew is difficult to say. But based on a three-foot jaw of one large specimen and analysis of the complete fossils of smaller specimens, it reached lengths of more than twenty feet and weighed more than three tons, larger than the biggest great white shark ever recorded.

Rhizodus was a ferocious predator with impressive hunting skills. With eight-inch fangs in both its upper and lower jaw—the name means "root tooth"—it had an extremely powerful bite and took large prey, probably including sharks. Like the arapaima today, *Rhizodus* was armored with tough but flexible scales, giving it protection while still enabling it to move quickly. Most likely it employed a vibration detection when hunting, ambushing its prey and using a "crocodile death roll" to kill and tear its victim to pieces. Like other rhizodonts, it could crawl onto land.

In other words, it was a terrifying creature with formidable physical attributes. In the end, however, it lost out. The world that followed the Devonian extinction favored the nimble, and in it the rhizodonts failed to diversify and adapt. They became, as the fish paleontologist Lauren Sallan puts it, "dead fish walking." And as Pangaea began to take shape, changing and drying up inland environments, the rhizodonts went out with a whimper.

When I was young, I saw Granddad at the Shedd Aquarium in Chicago. Granddad was an old Australian lungfish that had been brought from Australia by steamboat and train to dazzle attendees of the 1933 World's Fair, which was held in Chicago. The brown-spotted, four-foot-long lungfish certainly dazzled me, even though he spent much of his time at the bottom of the enclosure imitating a fallen tree log. Every now and then he would rise from his apparent apathy, slowly flap his large and fleshy pectoral and pelvic fins, and come to the surface for a slurp of air, letting out long, loud snorts as he breathed. I thought that was very cool.

No one had been able to establish Granddad's exact age, but the fish had been a mature adult when it was brought to the United States, so it was extremely old. I thought of my own grandfather, who was born in 1912, the same year that the *Titanic* sank. It was possible that Granddad had been born around the same time as my granddad, and I reflected on all that had happened in the world since then. What I didn't think about was how old the species was; that I was looking at a fish whose ancestors lived more than four hundred million years ago, making it the oldest living fish today.

Lungfishes belong to a group of lobe-finned fish called dipnoids, whose excellent fossil record stretches back to the early Devonian period. They emerged in marine waters, but soon moved into freshwater habitats, where they diversified, and by the late Devonian the rivers had become loaded with various lungfish species, including massive predators the size of small sailboats.

Over time, defying Cope's rule, they shrank in size, with smaller lungfish known as *Ceratodus* abundant during the Triassic era. Today, six species of lungfish remain. In addition to the Australian, or Queensland, lungfish, there are four African species of a different genus, which are thinner and have wispy fins, and a single lungfish species in South America that is even slimmer and resembles an eel.

One of the African species is the marbled lungfish, which is found in the Nile River and in lakes in East and Central Africa. It has a more striking appearance than the others, with a spotted, or marbled, body pattern that has earned it the name "leopard lungfish." It is also the largest lungfish, reaching lengths of more than six feet. This meant that the lungfish qualified for my list of megafish. Although I knew that the world's largest freshwater fish would not be a lungfish, my fascination with them evolved as I grew older and I began to understand their evolutionary importance.

Lungfishes have retained some remarkable adaptations from the time they first emerged. Like gars and arapaimas, they are able to breathe air, using one or a pair of lungs as a modified swim bladder that absorbs oxygen and removes wastes. When breathing, lungfishes will swim upward and position their heads so that the tip of the snout barely touches the water's surface. The mouth is then opened wide, and the fish sucks in air from just above the water.

The African and South American species use their long, spaghetti-like fins to touch and sense their surroundings. They are highly sensitive to pressure and turbulence and have a very good sense of smell and taste, making up for the fact that the lungfishes are almost blind.

Then there is the ability of African lungfishes to survive in a state of estivation, similar to hibernation. To prepare for droughts, they burrow into the bottom of a river or lake bed and encase themselves in a mucous sheath that gradually hardens. There they will stay, as the waters dry up, even digesting their own muscle tissue to obtain nutrients. African lungfishes dig in and encyst in this manner even if there is sufficient time to swim to deeper waters, and they can stay in such an induced state for more than two years.

But perhaps the most extraordinary thing about lungfishes is that they may have been the first animals to walk. We've long known that lungfishes are close relatives of early tetrapods, the four-limbed animals who were the first to move from water to land and thought of as the first "walkers." As a 2011 study in the journal *Science* showed, however, the evolutionary route to land-walking may have actually begun with lobe-finned fishes, like the lungfishes.

Filming West African lungfishes in lab tanks and analyzing their movements on computers, evolutionary biologist Heather King and her colleagues at the University of Chicago found that the fish moved around the tank by pushing off the bottom with their pelvic fins, which correspond to hind legs, alternating between fins in a walklike pattern and sometimes switching to a bounding, synchronous gait. With each step, the lungfishes lifted their bodies up and forward, much like some tetrapods do when walking. The 2011 study was the first time walking behavior had been shown in a fish related to the first land-walkers, and it indicated that the ability of animals to walk actually originated underwater.

The findings also cast doubt on fossil tracks previously credited to early land tetrapods with feet or toelike digits, such as those discovered in a fossil deposit in Poland in 2010 and attributed to a 395-million-year-old land creature. Those "footprints," scientists reasoned, were more likely to have been left by lobe-finned fishes moving along the waterbed, with some of the patterns shown in the fossil trackways similar to the patterns made by the limbs in lungfish.

The more I learned about ancient fish like the lungfishes, the greater grew my wonder at their evolutionary history. Hundreds of millions of years ago, these fish had developed extraordinary physical characteristics and behaviors, like the ability to breathe air and to bury themselves in the mud to escape drought. They may have even paved the path for animals to walk, and the marbled lungfish actually has the largest known genome of any vertebrate.

Yet lungfishes and other ancient fish are often referred to as "primitive." This makes no sense to me. When used scientifically, the term may suggest that an organism developed its features early on. Using the term more generally, however, implies that such an animal is not as evolved as other, more recently developed creatures, which is not true. In fact, every organism is the most evolved of its kind. As Allyse Ferrara says about the gars: "They got it right early on." And so did the lungfishes. In the evolutionary game that has wiped out more than 99 percent of all vertebrate species through time, they are indisputable winners, evolved for survival.

One of the first *Monster Fish* shows that I filmed was an episode about the largetooth sawfish. With a sharklike body and long snout—known as a rostrum and lined with sharp transverse teeth arranged to resemble a saw—the sawfish is one of the craziest-looking fish you'll ever find. In our pitch to National Geographic television, we described it as "a cross between a shark and a chainsaw."

The largest of five sawfish species, the largetooth sawfish (*Pristis pristis*) is also known as the common sawfish, because it was once found around the world. Today, however, that name is a bit of a misnomer, since the common sawfish, like the other sawfish species, is among the most critically endangered fish on Earth.

The Fitzroy River in Australia had become one of the world's last strongholds for the largetooth sawfish, and in 2007 our film crew flew into Derby in the northwestern part of the country to do a show about this peculiar fish. There, we met up with Dave Morgan, a sawfish expert at Perth's Murdoch University. The plan was for Dave to guide us on a ten-day trip along the Fitzroy River, one of the last unregulated rivers remaining in Australia, rising in the Durack Range in the east and running 325 miles southwest into the Indian Ocean. We'd camp at a string of sites that Dave regularly used for sampling sawfish and film as we went along.

We arrived during the dry season, and it was blisteringly hot. The Kimberley region, through which the Fitzroy runs, routinely gets up to 115 degrees Fahrenheit. Long inhabited by Aboriginal people, who settled here forty thousand years ago, the region is sparsely populated, even by

Australian standards, with fewer than fifty thousand people living in an area larger than California. We began our trip by driving 160 miles to a small settlement called Fitzroy Crossing, which Dave had described as "a two-pub town." I don't remember passing many cars as we drove through the harsh landscape, red dust from the parched dirt road kicking up behind us.

It seemed strange to think that megafish could live in such a desert world. Yet the Kimberley region is a biodiversity hotspot, largely because of the Fitzroy River, which has twenty tributaries and a huge catchment area, the nearest river running four hundred miles to the south. This means that when the monsoon season hits, all the falling water runs into the Fitzroy, transforming it into a raging torrent. Conversely, during the dry season, the river recedes to secluded pools and wetland refuges.

This kind of flood-pulse system has an explosive effect on biodiversity, as I was well aware from the Mekong, and the Fitzroy is no exception. In addition to two species of sawfish—the largetooth and the dwarf—its aquatic habitat supports more than three dozen species of fish, with many of them endemic to the Kimberley region. Many of Australia's well-known native animals, from the kangaroo to the kookaburra, are found here, with the lands and waterholes flanking the river also providing critical habitat for waterbirds and many endangered turtles and frogs.

Compared to lungfishes, sawfishes are relative newcomers on Earth. But they are still ancient, appearing in the fossil record about 60 million years ago, soon after the dinosaur-killing Cretaceous extinction event. Belonging to the elasmobranchs, they are actually a type of ray. Of the five species living today, the largetooth sawfish is the largest and can grow to more than seven meters, or twenty-three feet, which would make it a contender for the world's largest freshwater fish. Except it's not a true freshwater fish, as it spends much of its life in marine waters.

Central to the sawfish's evolutionary success is the rostrum, which is evolved not only for defense but also for hunting. Covered with electro-sensitive pores, the rostrum allows the sawfish to detect slight movements of prey hiding in the mud, and it can be used as a digging tool to unearth buried crustaceans. When prey try to swim by it, the usually lethargic sawfish will spring from the bottom and slash at the prey with its saw, stunning or impaling it sufficiently for the sawfish to devour it whole.

But the same physical feature that has proven such a useful tool for survival has also made the sawfish susceptible to the ravages of humans—more specifically, human fishing nets, which can easily ensnare the sawfish and its toothy snout. And while incidental catches in fisheries are a major problem, the sawfish has also been heavily targeted as both a trophy and food fish, with disastrous results.

Historically spanning the tropical and subtropical waters of all the oceans in large numbers, populations of the five sawfish species are estimated to have declined by 95 percent. The largetooth sawfish has been extirpated from Mediterranean and European waters, and it has not been seen in the United States since 1961. Disturbingly, this decline occurred over many years with few conservation efforts made to stop it. It was not until 2007, the same year that I traveled to the Fitzroy, that international commerce of sawfishes was banned globally.

In Australia, too, the sawfishes had disappeared from much of their historical range, and even in the Fitzroy River the largetooth was not safe. Killing them was outlawed, but not everyone followed the rules, as evidenced by the number of pubs around western Australia that displayed sawfish rostrums as trophies on their walls. A newspaper report that I read while I was there said a fisherman had caught two hundred sawfishes in his nets, many of them juveniles. Instead of disentangling the fish from his nets, the man had simply broken off the rostrums of the fish and thrown them back in the water to die. Dave said such stories were common.

From Fitzroy Crossing, we moved farther upstream, past Geikie Gorge, a one-hundred-foot deep canyon carved by the river into the remains of an ancient limestone barrier reef that formed in the Devonian period. At this time of year, the parched river was running so low it seemed like it wasn't flowing at all, and in some places it had been cut off almost completely, creating isolated pools of water that stretched a few hundred yards in length. The sites where we fished were all situated along deeper sections of the river, and we camped on the riverbank.

Joining our film crew was Dave, an American graduate student of his, and a team of Aboriginal rangers from a local community known as Jarlmadangah, with whom Dave had worked for many years. Aboriginal people are referred to as the "traditional owners" of the Fitzroy River, which goes by several Aboriginal names, including Raparapa. Yet much of the land in the Kimberley region is designated as pastoral and used for commercial cattle ranching.

A conflict over land use has long pitted the indigenous communities against the commercial ranchers, who have sought to divert water from the Fitzroy River for farming operations. Working in coalition with environmental organizations, Aboriginal groups have blocked numerous attempts to dam the Fitzroy over the years, and in the early 2000s such a coalition defeated a government proposal to send water by canal from the Fitzroy to drought-stricken Perth, one thousand miles to the south.

For the migratory sawfish, it was essential that the Fitzroy remained free-flowing. As scientists had relatively recently established, largetooth sawfish are born in estuaries before migrating upstream to spend their first four or five years in the river. On nearing maturity, they return to coastal waters, where they continue to grow. Blocking the river would clearly threaten the survival of the sawfish.

We fished for the sawfish using a gill net that had large mesh, so as not to entangle the fish too much. Dave placed it in a deeper section of the river where he had caught sawfish before, with the net reaching down to the river bottom, which is where the sawfishes, like other rays, spend much of their time. Because its gills are located on its underside, where water does not easily flow over them, the sawfish, like other rays, instead draws in water through large holes located behind the eyes, called spiracles, to breathe.

After putting the net in, we returned to the bone-dry riverbank. But we didn't want to leave the net in the water for too long. If a sawfish got caught, it was important to get it out as soon as possible. So, an hour later, we returned to check on the net. As soon as we began pulling it up, we could feel that something was stuck in it. It turned out to be a sawfish, and an impressive specimen at that, measuring seven feet long.

The fish had probably not been sitting in the net for very long, yet it had already managed to get its rostrum ensnared pretty badly. Luckily, Dave knew what to do. He turned the fish over on its back, which put it in a sort of catatonic state. Then, using a screwdriver, he methodically pried the net loose from the fish's rostral teeth, creating space between the teeth and the net as he went along. Soon, the net came off completely, and we took measurements.

Handling a large fish is always a thrill. Immediately, I am filled with adrenaline, trying to figure things out. How does the fish feel? How does it move? How does it react to being caught? Or being handled? Is it injured? How can I ensure that the fish is released back into the wild unharmed? It's my favorite part of the job, being in the field, handling fish as rare and special as the sawfish.

But I also immediately thought of how easy it had been to catch the sawfish. It didn't have a chance against the net, even one with a large mesh. The fish's awkward rostrum was always going to get it ensnared. The same novel adaptation that had yielded such evolutionary success had become maladaptive in the face of exploitation by humans and their fishing nets.

Like the lungfishes, the sawfish had evolved for survival. Now it seemed to have evolved for extinction.

Over the following days, we made our way down the Fitzroy, sometimes driving where there were no roads, setting up camp for the night on dusty, red soil, and sleeping under giant gum trees, as the locals call eucalyptus. We continued to catch sawfish, most of them about the same size as the first one, about two meters long, which meant they were young, not yet reproductively mature. The larger adults had migrated into the sea, where they would stay their whole lives. Dave had never seen a mature largetooth sawfish in the river proper.

Oftentimes, the nets would also hold a surprise catch: crocodiles. Freshwater crocodiles—or "freshies," as the Aussies call them—are common all along the Fitzroy River, and almost as susceptible as the sawfish to getting caught in nets. They're not particularly dangerous to people, as they will not attack unprovoked, and most of the freshies we caught were juveniles, about three or four feet long.

But on a couple of occasions it happened that we caught a six-footer in the net, and disentangling a crocodile of that size required some experience. Of course, Dave and the rangers all knew how to do it, careful to hold on to the freshie's head and not let it whip around to get a bite in. I learned the technique as well, though preferred to handle smaller individuals, in the one- to two-foot range.

A bigger problem as we approached the mouth of the river was saltwater crocodiles. Unlike the freshies, "salties" can be aggressive and have a reputation for being dangerous to people, with attacks each year ending in fatalities. Opportunistic predators, salties can lurk patiently beneath the surface near the water's edge, waiting to explode from the water with a thrash of their powerful tails to snatch unsuspecting prey, including water buffalo, monkeys, wild boar—and people.

Salties also presented a big threat to the sawfish. Many of the sawfishes that Dave had tagged displayed bite marks from encounters with crocodiles, and also from bull sharks. Some years later, he did a study showing that 60 percent of sawfishes analyzed had been attacked by crocs or bull sharks.

Down in the estuary, we caught another species of sawfish called the dwarf sawfish (*Pristis clavata*), which grows only about four feet long as an adult. With the estuary being a pupping ground for the largetooth sawfish, the largetooths that we caught there were also small, seemingly newborns, with the teeth of their rostrums still in the early developmental stage.

As I had learned, all sawfishes are ovoviviparous, a method of reproduction whereby the young develop inside a weakly formed egg shell within the adult female, receiving nourishment from their yolk sac. During development, the sawfish's rostrum is soft and flexible and the teeth are enclosed in a sheath. This, it seems, is another ingenious evolutionary adaptation of the sawfish, allowing the female sawfish protection from a sharp rostrum while giving birth.

Incidentally, a few years ago, researchers in Florida made another astonishing discovery about the sawfish's reproduction: that a small proportion of females of the smalltooth sawfish (*Pristis pectinata*), a species that maintains a relatively healthy population in Florida waters, reproduce on their own, without any male input, through a process known as parthenogenesis. Such asexual reproduction had been observed in various sharks, snakes, and fish in captivity. But until that discovery, the so-called virgin births had never been observed in vertebrates in the wild.

I left Australia in amazement of the sawfish. But also with apprehension for its future, which in the case of the largetooth sawfish had become inextricably linked to the future of the few rivers in the world that it could truly call home. A few years later, in 2011, the Fitzroy River system was placed on the Australia National Heritage List for its outstanding cultural and natural values. But the battle over it has continued, with commercial ranchers pushing for irrigation schemes that could "supercharge" the region's cattle industry, using water that some of them say would otherwise run "uselessly" into the ocean.

There is one other group of ancient fish that could be considered the most endangered animals on Earth: sturgeons. Out of the twenty-seven species of sturgeons, sixteen are classified as critically endangered on the International Union for Conservation of Nature (IUCN) Red List of Threatened Species, with four possibly extinct.

Sturgeons are one of the oldest families of bony fish in existence, with a story that dates back to the Middle Jurassic era, some 170 million years ago. They likely evolved from a long extinct group of ray-finned fish called palaeonisciforms, which emerged at the end of the Silurian period, more than 400 million years ago.

Today, most sturgeon species occur in the Ponto-Caspian region of western Asia, which includes the Black and Caspian Seas, though about a third are also found in North America. While most species are anadromous bottom-feeders, there are some sturgeon species that spend their entire time in fresh water.

Sturgeons are distinctive for their appearance: elongated bodies, lack of scales, and often great size. They have bony plates called scutes that cover the head and five rows of similar plates along the body. The toothless mouth, on the underside of the snout, is preceded by four sensitive tactile barbels that the fish drags over the bottom in search of prey.

With their prehistoric look, sturgeons are often referred to as "living fossils," a term coined by Charles Darwin in *On the Origin of Species* in 1859 to describe animals (platypuses, lungfishes, and sturgeons) that appear to have endured to the present day without undergoing much change. Like the term "primitive," "living fossils" carries a negative connotation, as if these fish have been unable to keep up with the times. But as we've seen with lungfish and gar, getting it right evolutionarily from the start should hardly be considered a disadvantage.

In any case, it appears good old Charles may have been wrong about the sturgeon. For a study published in *Nature* in 2013, researchers at UCLA analyzed the evolutionary relationships between nearly eight thousand species of fish and found that in at least one measure of evolutionary change—that of body size over time—sturgeons have actually been one of the fastest-evolving fish on the planet, with many of the species growing to very large proportions.

The largest of them all is the beluga sturgeon (*Huso huso*), which has historically been found primarily in the world's largest inland body of water, the Caspian Sea, from which it can travel up any of the more than one hundred rivers feeding it.

Many organizations, including the US Geological Survey, refer to the beluga as the biggest freshwater fish in the world. But as mentioned earlier, the beluga is not strictly a freshwater fish; it is capable of moving freely between fresh water and estuaries. It can make migrations of more than six hundred miles up rivers to spawn, but it also spends a lot of its time in the Caspian's salt water.

Be that as it may, I've always taken a keen interest in the beluga and other sturgeons, even those that aren't strictly freshwater fish. Not only is the beluga a giant, it's also one of the most long-lived of all vertebrates, with reports in the past of fish reaching ages over one hundred years. But you'll be hard-pressed to find any beluga sturgeons that old anymore—or, for that matter, any as big as they once grew. That's because, for centuries, fishers have been hunting the belugas and killing them as soon as they reach sexual maturity, to get their hands on the "black pearls" that the females possess, unfertilized eggs also known as caviar.

Throughout history, caviar has been seen as a food delicacy without parallel. Ancient Greek, Roman, and Chinese literature all refer to sturgeons and caviar, and the Chinese reportedly traded in the stuff as early as the tenth century. But it was not until the late twentieth century that the coveted eggs truly became the jewels of the luxury food scene and global demand for them skyrocketed. In 1999, the market price for the most sought-after caviar, beluga, shot up to $4,500 a pound, according to the Guinness World Records.

To satisfy such demand, belugas were intensely harvested, which caused populations to crash. Fishing regulations that existed were widely ignored, with harvesting through poaching in the Volga-Caspian basin alone estimated to be ten to twelve times over the legal limits. At the turn of the millennium, poachers there reportedly removed half of the mature beluga individuals each year.

Efforts to combat the beluga annihilation were ramped up in the 2000s when scientists launched the World Sturgeon Conservation Society and an international campaign to save the belugas. In 2006, CITES, the Convention on International Trade in Endangered Species, which had overseen the trade in caviar since the late 1990s, suspended all trade made with the traditional caviar-producing regions of the Caspian and Black Seas, while the US Fish & Wildlife Service banned the importation of beluga caviar originating from that region into the United States.

By that time, sturgeon farms had long been established in many places, including Russia, Iran, and even the United States, and global aquaculture production of beluga caviar far eclipsed that drawn from wild fish. But with demand still significantly exceeding available supply, illegal fishing continued and legal fishers were forced to catch increasingly younger and smaller fish. Meanwhile, the high prices that eggs from wild-caught fish attracted enticed some traders of farm-raised sturgeon to label their caviar as if it came from wild fish to bump up its value to consumers, a practice dubbed as "black washing."

When the beluga sturgeon was finally classified as critically endangered on the IUCN list, in 2010, the recognition of its catastrophic decline seemed long overdue. It was hard to see how the trend could be reversed. The giants were disappearing, and it felt like there was nothing we could do about it. And if you don't think something can be done, then nothing will be done.

There was no getting around the fact that giant fish, which had been on Earth for hundreds of millions of years, were disappearing in a flash. The idea of the sawfish being "evolved for extinction" even suggested its demise was somehow preordained and that there was nothing we could do about it. But of course that was not true. The decline of giant fish was one hundred percent the result of human activity. And if humans were the ones causing this catastrophe, then humans could also stop it.

For all the disheartening developments, there were also positive things happening everywhere I looked. A fishing moratorium had been implemented for the Mekong giant catfish. Local communities along the Mekong and all over Southeast Asia were setting up fish reserves and taking action to protect their rivers. The same thing was happening in South America, Europe, and Africa. In Florida, a gill net ban was fueling a modest sawfish recovery. Scientists were discovering and describing hundreds of new species of fish, always pushing to learn more.

One species that had seemed to have no future was the alligator gar. Throughout the twentieth century, it had been denounced as nothing but "trash fish." Any attention paid to it was negative and attributable to misconceptions surrounding the species and its impact on sport fishes. Historic removal efforts, chronic overfishing, and habitat loss due to river fragmentation and flow modifications had resulted in the extirpation of gar populations from many drainages.

But something strange happened. The image of the alligator gar began to change. It started with a treatise written by a University of Idaho fisheries biologist named Dennis Scarnecchia, which was published in the journal *Fisheries* in 1992. Not only was the view of alligator gars as harmful to game fishes and recreational angling wrong, Scarnecchia wrote, but the giant fish were important contributors to ecosystem stability and function, balancing predators and prey, and therefore instrumental to more successful angling in the long term.

Until the late 1990s, no one had really studied the alligator gar. As a result, basic knowledge about its life history and ecology lagged far behind that of most fish. The little work that had been done had focused largely on the fish's diet; the long-held belief was that gars ate the game fish humans wanted to catch. As it turned out, they don't. Researchers like Alyse Ferrara and her colleague Solomon David, an ichthyologist from the same university, Nicholls State, found that gars eat more or less whatever prey species are abundant, and are more likely to go after shad and minnows—and even things like small mammals, waterfowl, insects, and crustaceans—than game fish, such as sunfish and bass. Scientists were able to develop a greater under-standing of the important role that alligator gars, as top predators, play in maintaining healthy ecosystems. The knowledge of the biology and ecology of the species has been crucial to advance science-based management.

Efforts to actively manage alligator gar populations have been successful in the Trinity River, the longest river with a watershed entirely within the state of Texas. In 2009, the Texas Parks and Wildlife Department instituted a statewide one-fish-per-day regulation for alligator gar, intended to maintain the current status of the Texas population until the department had amassed enough information to effectively manage stocks based on scientific data.

Since then, researchers have found that alligator gar populations in Texas are relatively healthy, both in abundance and size of fish. They've learned that females grow significantly larger and older, more than sixty years. Years with particularly high recruitment (the process by which very young, small fish survive to become older, larger fish) occur relatively infre-quently, and rely on the timing, magnitude, and duration of flood events. "All of this information has been critical in developing science-based rec-ommendations for the management and conservation of Texas alligator gar fisheries," says Dan Dougherty, a research biologist with Texas Parks and Wildlife's Heart of the Hills Fisheries Science Center.

After my time in Louisiana I traveled to the Trinity River, where a fishing guide named Kirk Kirkland introduced me to alligator gar angling, a grow-ing sport on the Trinity. People arrived from all over the world, Kirk said, to catch a river giant that was now being admired for its size and power. Visitors were almost guaranteed a catch in a few days of fishing, and likely a catch over a hundred pounds, maybe even over two hundred.

Seeing the healthy population of alligator gars on the Trinity under-scored how important it is to tell the good news—along with the bad—about our river giants. The alligator gar still had a long way to go, and there was still a lot we didn't know about this species. But saving it, and others like it, was no lost cause. And that was a story worth telling.

A River Runs Short

Few rivers on Earth have both changed the world through which they flow and had as much change brought upon them as the Colorado River. Following the creation of the Rocky Mountains beginning seventy-five million years ago, the Colorado first formed as a river flowing straight west. It was not until the rising Colorado Plateau reached its peak six million years ago that it established its present course, running southwest from the Rockies, through what is now the Grand Canyon, forming a vast marshy delta as it flowed into the Gulf of California.

Of course, the Grand Canyon didn't exist then. It was the powerful waters of the Colorado and its tributaries that cut through layers upon layers of rock to forge what is today one of the world's greatest natural wonders. For millions of years, the Colorado kept flowing, its power unchallenged, until eventually humans set foot in the river, and nothing would ever be the same again.

It was long thought that the first humans settling on the Colorado Plateau—the desert area that today centers on the Four Corners region, where Colorado, Utah, New Mexico, and Arizona meet—were Paleo-Indians of the Clovis and Folsom cultures, who arrived there around twelve thousand years ago. But recent discoveries of human settlements at an excavation site called Eagle Rock, near the small town of Delta, Colorado, have pushed that time line back at least a thousand years.

Over the following millennia, human use of the Colorado River basin revolved primarily around hunting and fishing. But with time various tribal societies began to garden and eventually developed large farms. Past indigenous groups are known to have farmed large swaths of land along tributaries in present-day Arizona more than a thousand years ago. Archaeologists have uncovered storage reservoirs and hundreds of miles of ancient canals around Phoenix, evidence of what may have been the largest irrigation system built by prehistoric peoples in North America, a remarkable feat considering canals would have been dug by hand using wood, bone, and stone tools.

After the United States purchased the territory of the Colorado River basin from Mexico in 1848, the discovery of gold in California triggered a western migration in search of riches. The importance of the Colorado to the future settlement of the southwestern United States became increasingly clear. With water, anything was possible in the desert, and as agriculture intensified throughout the basin, scholars referred to the river as "the Nile of America." Like the Nile, the Colorado originated in the mountains, flowed through a hot desert, and carried with it substantial amounts of silt down to the sea. But the expansion of agriculture required more than small-scale gravity irrigation systems. It was time to build bigger dams.

By the turn of the century, most dams that existed in the western United States were small private ventures, mostly constructed from earth and rock, diverting water for mining businesses or irrigation of personal properties. But technology emerged to build bigger concrete dams for larger-scale irrigation, water supply, and flood control, and a new federal program created by the Newlands Reclamation Act of 1902 paved the way for such dams to be built in the Colorado River basin. Within a few years the first one, the Laguna Diversion Dam, had been completed. Located on the main stem of the Colorado, near the Mexican border, the dam diverted water to the farm fields of Arizona's Yuma Valley. It also ended boat travel to the north.

Meanwhile, a much bigger dam was being constructed northeast of Phoenix, on the Salt River, a Colorado tributary. The Theodore Roosevelt Dam would store water and provide flood control at a huge scale. When it opened, in 1911, it was the largest masonry dam in the world—280 feet high—and for a while the Roosevelt Lake, created by the dam, was the world's largest artificial reservoir. The project cost a staggering $10 million, but was considered a huge economic and engineering success. Irrigating vast land areas, the Roosevelt Dam would contribute more than any other dam to the settlement of central Arizona.

California's farmers also coveted the precious waters of the Colorado, which marks the state line between Arizona and California but never enters the state proper. From early on, California sought to bend the Colorado to its own benefits, sometimes causing great ecological upheaval. One example is California's largest lake, the Salton Sea. It formed in 1905 when massive flooding caused the Colorado River to break through an irrigation canal headwork that had been built as part of a project to provide water to

California's Imperial Valley. For eighteen months, water flowed freely into the Salton basin, before the levee break was filled and the Colorado River forced back into its channel.

The newly formed Salton Sea created a rich oasis for birds, and the later introduction of various game fish built up a lucrative sport fishery around the lake. But without a natural outlet, water trapped more than two hundred feet below sea level evaporated over the years, leaving increasing levels of salt behind. The result has been an environmental disaster for birds, fish, and humans alike, as the shrinking Salton Sea has transformed into a salty dust bowl.

But it wasn't just farms in California that needed water to grow in the early 1900s; cities did too, especially fast-expanding Los Angeles. With water supply from its eponymous river woefully inadequate to sustain LA's unprecedented growth, planners in Southern California sought to harness water from far-flung places for their ever-growing desert cityscapes. In 1922, California had been one of seven states to enter the Colorado River Compact, an agreement to provide for the "equitable division and apportionment" of the use of waters from the Colorado River system, allowing California to stake a greater claim to the Colorado.

Eventually, Southern California secured its future water supply from the Colorado River with the construction of the Parker Dam on the California-Arizona border. Completed in 1938, the dam created Lake Havasu, from which water would be pumped through the new Colorado River Aqueduct running west across the Mojave Desert to the east side of the Santa Ana Mountains, and supplying almost all cities in the greater Los Angeles, San Bernardino, and San Diego areas. To this day, it is one of the primary sources of drinking water for Southern California.

By this time, a dam construction boom was well under way in the United States, with new federally funded projects designed not only for water supply and irrigation purposes but also for generating electricity. The biggest was the Boulder Dam, later renamed Hoover Dam, built on the Colorado River on the border between Nevada and Arizona, upstream from Lake Havasu. Completed in 1936, Hoover Dam became both the highest and largest dam in the world, and once online it earned the title of the world's largest hydroelectric facility.

For decades, building dams equated with progress. Few, if any, considerations were given to things like fish. But by the 1950s, a chorus of voices had begun warning about the environmental consequences of the dam boom. One project that came under particularly heavy criticism was the building of a huge dam at Glen Canyon near the town of Page in northern Arizona. Initial plans to build the dam in the 1920s had been scrapped, only to be resurrected some thirty years later. Critics questioned the dam's economic justifications, as well as the environmental damage caused by building a dam across a gorge lined with sandstone, which would flood the scenic Glen Canyon and threaten the integrity of the Grand Canyon just downstream.

Still, the plans went ahead, and in 1963 the Glen Canyon Dam stood completed 583 feet above the Colorado River, the second largest dam in the United States after Hoover, and with a newly plugged Lake Powell next to it. By then, all resemblance to the historic conditions in the Colorado River had long since faded, with every drop of water litigated and allocated for. In time, the mighty river that had once carved out America's most iconic landscape would no longer reach the sea.

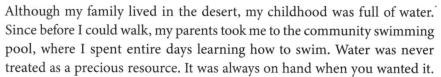

Although my family lived in the desert, my childhood was full of water. Since before I could walk, my parents took me to the community swimming pool, where I spent entire days learning how to swim. Water was never treated as a precious resource. It was always on hand when you wanted it.

In the backyard of our house—on College Avenue in Tempe, Arizona, a mile down the road from the university where my dad taught economics— there was a metal pole with a wheel on it, and every couple of weeks we would turn the wheel and water gushed out, flooding the entire backyard. For a whole day, my brothers and I splashed around in the water, collected by the berms around the yard, until the water eventually disappeared, and then we would go look for monsters or play hide-and-seek in our lush jungle of trees and bushes.

The whole town had an agricultural feel to it, with little ditches—three feet wide and made of dirt or cement—filled with water and running along the streets. There was never any real talk of water conservation at our house, except when it came to showers, which were to be short. But that was probably just to save on heating costs.

It was different once you got out of Tempe. Arizona rivers, with names like Gila and San Pedro, rarely had water in them; they were ghosts, imprints on the landscape of water courses that barely existed any more. I figured that was normal. In fact, I grew up right next to the Salt River, which I knew mostly as a dry riverbed, and which we would cross when we traveled from Tempe into neighboring Phoenix or Scottsdale.

But there was water in nature if you went to look for it. As my two brothers and I grew older, our parents would take us camping. Our favorite hiking trails were those coming off the Mogollon Rim, the slanting escarpment that stretches across north-central Arizona and serves as a boundary between two distinct worlds—the cool high country above it and the burning deserts below. We hiked the narrow canyons, along magical streams of crystal-clear water, swimming and cliff jumping where the streams were deep enough.

As a kid, I didn't know that water powered many cities in Arizona. It wasn't a thought I had when I flipped a switch and the lights came on. Until Arizona began producing energy from coal-fired power plants in 1971, a couple of years before I was born, the state had derived the vast majority of its electricity from hydropower. In the early 1980s, it was still getting about 40 percent of its energy from hydroelectric dams. One of those dams was the Glen Canyon Dam, and in the late spring of 1983, when I was ten years old, this dam made big news.

On June 2, 1983, authorities announced the earlier than normal release of water from Glen Canyon Dam. The Rocky Mountains had received record snowfall the prior winter, and a sudden warming trend had caused a huge surge in runoff, with the Colorado River carrying enormous amounts of water into Lake Powell, the reservoir created by the dam, filling it almost to capacity.

Fearing catastrophic damage to the dam's hydroelectric generators, officials began to release the maximum amount of water, about forty thousand cubic feet per second, through eight gigantic turbines in the dam's power plant. But as Lake Powell kept swelling, officials decided for the first time ever to open one of the dam's two spillways, huge tunnels bored into the sandstone on either side of the dam and running parallel to the river.

The next morning, engineers heard strange rumbling noises from deep inside the dam works, and chunks of concrete could be seen ejected into the river from the open left spillway. The water took on a reddish hue, suggesting the spillway's concrete lining had been breached and that the red sandstone around it was washing into the river.

The spillway was closed, and an inspection revealed a series of holes in the concrete walls gouged out by water. Engineers discovered a small bump in the concrete lining. This resulted in "cavitation," which occurs when water rushing down a smooth surface encounters a bump. As it flows over that bump, a vacuum is created, and when that vacuum breaks, it sends shockwaves into the surface, with potentially devastating consequences. Could the concrete failures lead to the collapse of the soft sandstone around the dam?

The water in Lake Powell kept rising, and it was now inches from spilling over the dam gates. In a desperate effort to prevent catastrophe, engineers rigged four-by-eight sheets of plywood to the top of the gates to extend their height. The job took less than a day, and miraculously it worked. The new flashboards leaked at the seams, but they were enough to hold the rising water back, buying officials some time.

Some, but not a lot. The amount of water flowing into Lake Powell kept increasing. Officials increased the height of the flashboard structure to eight feet and replaced the plywood with metal. In a last-ditch effort, they also opened both spillways, increasing the flow even through the damaged left tunnel to try to blast some of the debris from it. Boulders the size of cars belched from the spillway into the river, forcing the officials to dial back the release flow and leaving them with few options.

The runoff eventually began to subside on its own, and on July 15, more than a month into the crisis, the lake level finally peaked, just inches from the top of the new eight-foot-high metal flashboards. A week later, officials closed the ravaged spillways.

The events at Glen Canyon Dam would spark an intense and ongoing debate over the safety and viability of dams, as well as their environmental cost. This debate figures prominently in my work now, but was not something I paid attention to growing up.

The strongest memory I have of Arizona water politics in the 1980s instead has to do with the Central Arizona Project. This was a project to bring drinking water from the Colorado River to residents of the Phoenix-Tempe area and much of the state through a 336-mile-long aqueduct built from the south portion of Lake Havasu, the same lake from which water supplying Southern California was drawn. Construction on the project had begun in 1973, but it would take until the late 1980s before the water hit our municipal delivery system.

The CAP, as the project was called, made an impression on me because the water tasted horrible. Water should be clean and clear. This stuff came from the tap warm and tasted of chlorine and salt. As it turns out, the river water had a different mineral mixture and flow pattern than the aquifer water, which caused it to stir up and dislodge rust in city water mains and house pipes. Everyone talked about the terrible water. Some cities even reverted to groundwater. But for most people, there was little they could do except install water filters at home.

By this time I was in high school, and that's when my interest in fish and things aquatic really took shape. After my freshman year, I spent the summer doing an internship at the National Marine Fisheries Service aquarium in Woods Hole, Massachusetts, the oldest public marine aquarium in the United States. There, I fed the fish and cleaned tanks. I also got to collect jellyfish, crabs, and other tide-pool creatures from the rocky beach adjacent to the aquarium. One time I mistakenly put the animals I had collected into what I thought was an empty tank, only to return the next day to find that it was actually a large anemone display. All the crabs and fish I had introduced the day before had been devoured by the anemones.

After a while, with the support of the generous staff members, I started giving short presentations to the public, and I helped with the seal show featuring two rescue seals, Skeezix and Cecil. In front of a crowd, I tried to get the seals to do simple tricks in return for a mackerel. Some days went well and the crowd was very enthusiastic, but a lot of days we couldn't get the seals to do anything.

I had always loved visiting aquariums. Seeing the lungfish at the Shedd, big catfish at the Steinhart Aquarium in San Francisco, and sharks and tuna in Monterey Bay all made an impression on me. But the internship in Woods Hole made me realize that I could do something that I truly enjoyed as a vocation.

When the time came to start college, at the University of Arizona in Tucson, I decided to study ecology and evolutionary biology. The major appealed to me because of its environmental and organismal focus, and also because the department offered field courses in conservation biology, ichthyology, and the natural history of the Sonoran Desert and the Gulf of California, which included travel to Mexico. More than anything, I wanted to get into the field.

In addition to my classes, I applied to the school's newly created Undergraduate Biology Research Program, which provided funding for undergraduates to gain summer research experiences. I received a position as a research assistant with the Arizona Cooperative Fish and Wildlife Research Unit, conducting surveys of native fish in the Colorado River. It was an experience that would change my life.

Few rivers in the world have seen their native fish fauna as decimated as the Colorado, which, of all the major rivers in North America, has always had the lowest diversity of native fish, particularly in the lower part of the basin (below the Glen Canyon Dam). Only eight fish species have evolved in the lower basin, and those belong to only two families: minnows and suckers, both of the Cypriniformes order. (In comparison, there are almost three hundred native fish species in the Mississippi River.) At the same time, the Colorado River has the highest level of fish endemism of any river system in North America, with six of the lower basin's eight native species found in no other river in the world.

Why is that? A lot of it has to do with the geographic isolation of the Colorado system, and also the harsh environmental conditions that existed before the river was transformed by humans. For sunfish, catfish, and other fish that evolved outside the basin, the Colorado's turbid and unpredictable waters, running through muscular mountains and scorching deserts, were likely too much to handle.

Drought no doubt played a key role in the evolutionary adaptations of fish in the Colorado. The late Wendell L. Minckley, an ichthyologist at Arizona State University who had five fish species named after him, speculated that seasonal low flows, amplified by drought, caused fish to retreat from broader river channels into deeper and more secure canyon reaches. Stuck in those spaces, some species would be eaten by others, and generally only a few larger fish would survive.

One such fish is the Colorado pikeminnow (*Ptychocheilus lucius*), which is believed to have evolved more than three million years ago. The only large fish found in the Colorado River system that, historically, reached lengths of up to six feet, this beautiful fish, with its long, streamlined body of olive-green hue, no longer grows this big. (Nor can it be found anymore in the Lower Colorado River basin.) With its big fins and small eyes, and a large mouth that, despite it being a voracious carnivore, lacks what we normally call teeth, the Colorado pikeminnow is the largest of four pikeminnow species and is known for long-distance spawning migrations of more than two hundred miles in late spring and early summer.

Like some other native Colorado fishes, such as the razorback sucker and bonytail, Colorado pikeminnows can live nearly fifty years and produce tens of thousands of eggs each spawning season. This has allowed the pikeminnows to survive through prolonged periods of drought when spawning may have been impossible. Once favorable conditions return, they can repopulate the river in a single season, in contrast to many other freshwater fish that live less than ten years and can produce only a few hundred offspring each year.

But the story of the Colorado pikeminnow is ultimately one of decline. Naturally, this has occurred at the hands of humans, and this time we've managed to cause it in a trifecta of progressively harmful ways.

From the time humans first arrived at the Colorado River, the pike-minnow has likely been targeted for food. Fishing practices of many Native American tribes throughout the years have been well documented. One nomadic people, the Chemehuevie, even derives its name from a Mohave word that means "people who fish." The Chemehuevie, who roamed southern Nevada and western Arizona, called the Colorado pikeminnow Ah´chee Ah´had, or simply "the best fish."

The Colorado pikeminnow was also a valued food source among the early white settlers, who alternatively called it "squaw fish"—a term commonly used until very recently, despite its blatantly derogatory connotation—and "white salmon," a reference to the pikeminnow's migratory behavior. There are plenty of stories of pikeminnows weighing up to a hundred pounds, with settlers describing catches as a harvest producing "quite a few meals for the family."

Historical accounts from the turn of the twentieth century tell of farmers diverting the river's water into their orchards; and after the water dried up, hundreds of pikeminnow being scattered across the fields, from where they were pitchforked out and used as fertilizer. Still, the pikeminnow remained the Colorado River's top predator in the early 1900s, and they were a popular target fish for anglers, known to take bait in the form of mice, birds, and even small rabbits, despite their "teeth" being found only on a bony circular structure located deep within their throats.

But people also wanted to fish other species, both for food and for recreation, and this brings us to the second way we have caused the decline of the pikeminnow: by introducing non-native fish species into the Colorado River. Stocking of non-native fish in the Colorado probably began as early as 1881, as the countryside was still being settled. By 1910, common carp, bullhead, and channel catfish were prevalent throughout the lower river.

Non-native species that take root in new ecosystems and start causing trouble are known as invasive organisms. In the case of the Colorado, the new arrivals increased the competition for food among all aquatic inhabitants. But even as food themselves, the catfishes cause great harm to the pikeminnows that prey on them. The pikeminnows get their prey's barbels stuck in their throat when they try to swallow the catfish, causing them to suffocate and die.

Even so, the pikeminnow might have survived these first two threats had it not been for the third and most destructive way we have caused the decline of the giant fish: by building dams. In the Colorado River system, the construction of dams has proved disastrous for the migratory pikeminnow;

neither the reservoirs behind the dams nor the cold, clear water flowing from them provided suitable habitat for a species that had evolved in wildly different and more extreme environments.

The last Colorado pikeminnow below the Glen Canyon Dam was recorded in 1975. The following year the species became federally protected and listed as endangered. Efforts to reintroduce the pikeminnow to the lower basin, using brood stock held at the Dexter National Fish Hatchery in New Mexico, failed because the fish would not reproduce, and eventually the efforts were abandoned. It left the Lower Colorado with the dubious distinction of being among the few major rivers in the world with an almost entirely introduced fish fauna.

It was not until many years later that it dawned on me that the Salt River, which had once flowed near my childhood house and at that time was a dry riverbed, had once held these monster fish.

***

While the Colorado River has been systematically dismembered, many of the smaller tributaries that feed into it in the lower basin exist more or less in their natural state. They were the focus of our University of Arizona research project. The goal was to map these tributaries to better understand if they could provide suitable reintroduction sites for native species that had disappeared from the main river.

Four teams, each with a master's student as team leader and an undergraduate, such as myself, as the assistant, studied four different watersheds. I got paired with Steve Weiss, an East Coast native of part Austrian background. An avid fly fisher, Steve had been passionate about freshwater fish since a young age, making him an outlier during his undergraduate studies at New Jersey's Rutgers University, where everyone in his program focused on the ocean.

I was going to assist Steve with his work on the Paria River, a tributary that originates from a series of springs inside Bryce Canyon National Park in southern Utah and runs for ninety-five miles before joining the Colorado in Arizona, fifteen miles below the Glen Canyon Dam. The landscape through which this river flows is awe-inspiring, especially where it cuts through the Navajo sandstone to create spectacular slot canyons like the Paria Narrows and the nearby Buckskin Gulch, one of the most popular spots for canyoneering in the American West.

Starting my freshman year, Steve and I frequently drove eight hours from Tucson to Lees Ferry, a fishing area and river rafting launch site where we'd set up camp. From there, we hiked up the river each day. Steve says he has seldom seen the enthusiasm and energy that I showed as I marched into the canyon, a bandana wrapped around my head for protection against the scorching sun, lacking in anything resembling experience but eager to learn.

While the scenery around it is spectacular, the Paria River itself is less impressive. In many places the water runs less than a foot deep, and everywhere it's extremely turbid, closely resembling the pre-dam Colorado River in its water chemistry. This makes it essentially unfit for any but the most mud-tolerant species—fish that, during floods, can withstand being blasted by suspended sand, which causes tremendous abrasion to their bodies and gills.

Our mapping efforts were concentrated on the lowest part of the Paria, from its confluence with the Colorado and ten kilometers upstream. Steve and I dragged seine nets across the river at increments of five hundred meters and logged whatever fish we caught. For the first several visits to the Paria, we didn't find any fish in the river, except one tiny species called the speckled dace (*Rhinichthys osculus*).

The speckled dace, of which there are fifteen recognized subspecies, is a member of the minnow family. Unlike the Colorado pikeminnow, its elongated body grows all of three inches long. And yet it is a tough fish and worthy of appreciation, even from someone who would end up studying fish more than ten thousand times more massive than those tiny dace. We found hundreds of speckled dace crowded in small, ephemeral mud pools no larger than a kitchen sink. It seemed outrageous that a fish could survive in such conditions, but the dace in these miniscule spring-fed pools seemed to do just fine.

The dace was a cool fish, but it was frustrating that our search for more river life came up short. While my enthusiasm never waned, it was tough work. On the lower part of the Paria the landscape was open, and it was hard to avoid the blasting heat, to say nothing of the scorpions and rattlesnakes.

Winter came and almost went when, one day in March, everything changed. We had been working on the lowest part of the river, less than a mile from its confluence with the Colorado, dragging a net across the knee-deep waterway to see if we could catch any fish. The river was running like thick chocolate milk, from rain or snowmelt, and we were covered in mud, when, suddenly, the water in front of us erupted in a boiling frenzy. There were fish everywhere, and definitely not dace.

These were flannelmouth suckers (*Catostomus latipinnis*), the only endemic "big river" fish native to the Colorado that is still commonly found throughout the lower section of the basin. We had been looking for signs of flannelmouths spawning in the Paria, because the suckers seek out tributaries or more shallow areas for their reproduction. But we thought we'd find larvae, not masses of adult flannelmouths actually spawning before our eyes, whipping and whirling out of the water.

Among the twenty-eight identified species of suckers, which are all native to North America, the flannelmouth is the largest and can grow more than two feet long and weigh about eight pounds. Although they have fared better than other endemic Colorado River fish, flannelmouths have also seen their numbers reduced in the lower basin. In the upper basin, meanwhile, the flannelmouth has hybridized with another sucker species, causing genetic degradation. These developments had just prompted the US Fish and Wildlife Service to categorize the flannelmouth sucker as a "species of concern," which is why finding the spawning school was so thrilling.

We immediately set to work collecting fish, measuring their lengths and weights, and tagging them so that their movements could be tracked if the fish were found again. Handling the fish was physically difficult. The flannelmouths were big, strong, and slippery, and we were covered in mud and fish slime. But I didn't mind. I was excited that we had found these fish in their natural state, presumably behaving the same way they have been for thousands of years, packed tight into a small, turbid river for their annual spawning run. This behavior mimicked what had once occurred downstream in the Colorado River but was no longer possible due to the changes brought about by the dams.

The fish stayed in the same spot for a couple of days and then disappeared as abruptly as they had emerged. By then, however, it was clear that Steve had found a subject to focus his research on: the flannelmouth suckers and their spawning in the Paria River.

The research that the different teams did individually was only one part of the larger project. As a group, we also carried out a general survey of fish along the main Colorado River and its tributaries. In large rafts, we traveled down the Colorado for eighteen days, which sounds like a spectacular adventure. And it was, except we did it in the middle of the summer, with daytime temperatures reaching 110 degrees Fahrenheit. The only way to escape the heat was to jump in the river. But as anyone who's been on the Colorado River knows, its water is extremely cold, since dams chill the water down to just 39 degrees. The rafting was challenging too; the Colorado River has some of the fiercest rapids in the world, despite the dams.

As we moved down the river, into the Grand Canyon and through some of the most stunning scenery on Earth, we stopped at tributaries and hiked up several miles, sweeping nets through the water and recording what we found, creating a map of where fish were living. At night we set up camp along the shore, and Steve inevitably busted out his guitar and sang Leonard Cohen songs. It was magic.

As we continued our work on the Paria River, Arizona entered monsoon season. In late summer sudden thunderstorms were common, and you had to be careful not to get stuck in flash floods inside the canyons. Theoretically, I knew how deadly these flash floods could be; I had heard stories of people getting swept up in them, often with fatal outcomes. Even the surveyor who, in 1922, had first gone looking for the ideal spot to build the Glen Canyon Dam had died in one.

Even so, as Steve and I returned to the Paria for yet another survey, this time with another student, there was little reason to believe we would run into any problems. The forecast suggested continued clear weather as we set off into the canyon from the White House trailhead. The canyon deepened and narrowed until we entered the Paria Narrows, where the towering sandstone walls reach eight hundred feet high and are so close to each other they leave a gap merely six feet wide. Eventually we reached the junction to Buckskin Gulch. We set up camp a few hundred yards up the gulch, on a small and sandy bluff.

We had just begun preparing for dinner when we heard a strange noise. It started quietly, like a soft breeze through aspen, but quickly built into a howl that sounded like the summer sandstorms that ripped through our backyard when I was a kid. None of this made any sense because the sky was clear and the air was still. The three of us looked at each other, confused and trying to understand what was happening.

I ran down the gulch in the direction of the Paria to investigate what was happening. I hadn't made it very far when Steve yelled at me to come back. As I looked down the gulch, I could see why. Water was coming toward me, up the gulch. I quickly returned to camp. There we stood, watching the water flood the gulch. Soon it was all the way up to the sand bluff, surrounding our camp. Luckily, it didn't go any farther. We were safe.

It was clear what had happened. A flash flood had swept down the Paria, and the floodwaters were so strong they backed up into the gulch. We had been extremely lucky to be out of the main canyon. I have heard since that the leading edge of flash floods on the Paria are not walls of water but walls of flowing mud and sticks that would have swept us up and carried us downstream, almost certainly to our deaths.

But we still had a problem. With the waters and mud piled up into the gulch, there was no way to get out. We stayed the night and hoped that the waters would recede the following day. But when we woke up the waters were still high. We made several attempts to hike out through the main canyon. But not only was the water still too high, it was also bitterly cold, so we returned to camp. Our food supplies were dwindling, but we had no choice but to stay another night.

The next day, the water levels sank to about waist deep, and we decided we would make another go of it. With our heavy backpacks, we made it into the canyon. But we had miscalculated the depth of the water, the cold, and the difficulty of hiking upstream against the current. At several spots the water was so deep that my feet couldn't get traction on the river bottom, and the river carried me back downstream. Steve, who was taller, had to grab me and get me through the deepest sections. Above us, we noticed the high-water mark on the sides of the canyon walls, ten feet above our heads, and were reminded that should another flood come down the canyon now, we would have no chance.

Gradually the canyon opened up and the water level lowered. But now we encountered another problem: quicksand. The high water had moved and redistributed huge amounts of mud that looked like normal sand. But if you stepped in it, the top layer would collapse like the skin on a pudding, and you would sink up to your waist. Again, Steve had to help me navigate the thick mud.

Eventually we made it back to the trailhead and our car, muddy, cold, and exhausted, but happy to have made it out and with a new appreciation for the Paria and a reinforced respect for the power of flash floods. Never in my life had I been at the mercy of the environment like that. I realized there are still moments and places in this world where we are powerless to control nature.

Steve went on to write his thesis: "Spawning, Movement and Population Structure of Flannelmouth Sucker in the Paria River." While spawning occurred throughout the lower ten kilometers of the river, he found no evidence of juvenile flannelmouth rearing in the Paria and Glen Canyon area in the prior twelve years. Even for a fish that is doing relatively well, the dramatic changes to the Colorado have left the flannelmouth with few options. It is a fish evolved to survive in the natural conditions of a Colorado River that no longer exists.

After Steve finished his research, he moved to Austria, while I continued my work on the Paria River with another University of Arizona master's student, Michele Thieme. Building on Steve's findings, Michele focused on what the flannelmouth sucker was doing at the confluence of the Paria and the Colorado. Our research showed that in the spring, the still, deep water there was a staging area for flannelmouth preparing to spawn, and in the early summer, if the conditions were right, it provided habitat for newly hatched fish.

A fun part of the project was tagging adult flannelmouth suckers with acoustic transmitters. The tags, slightly larger than an AA battery, were inserted into the fish's body cavity and transmitted a coded signal that would enable us to identify the fish remotely, with either a stationary or mobile receiver, to track their movements. The work was in preparation for a massive experiment scheduled to occur later that year on the Colorado River: a controlled flood. Bruce Babbitt, the former Arizona governor and then US secretary of the interior, was coming to open the floodgates of Glen Canyon Dam to release a massive flood down the Colorado, an experiment designed to mimic the natural snow-melt flows that historically tore down the Colorado each spring.

On the day of the release, Michele and I sat in lawn chairs on the riverbank fifteen miles downstream from the Glen Canyon Dam, watching the water gradually creep up and eventually inundate the entire beach. Bigger and more powerful than I had ever seen it, the river carried large amounts of sand and wood downstream, redistributing sediment and nutrients throughout the canyon. It was a grand experiment. Even so, the flows were just a tenth of the highest flow that had been recorded on the Colorado—yet another reminder of how far from the historical conditions the river had been removed.

The flannelmouth suckers didn't seem to pay much attention. They retreated into the mouth of the Paria, which formed an even larger backwater during the flood. As Michele eventually showed in her thesis, however, increased flooding helped improve conditions for the flannelmouth suckers and was crucial to their reproductive success.

My experience on the Colorado River changed the way I looked at fish and humans and our place in the natural world. For the first time I understood that my love of water and aquatic life could be a career. I saw Steve and Michele, a few years ahead of me in school, pursue degrees that would enable them to work as wildlife or fisheries professionals. I saw them struggle with research questions; attempt to make sense of the patterns we observed in the field; and work to develop a research project that would provide answers to previously unanswered questions. One of my mentors at the university, Richard Kissling, described this process and the feelings associated with it as the "spirit of inquiry," and I identified with that spirit.

From Steve, I learned about endangered species issues; of the eight species of fish once found in the Lower Colorado River, three were gone and two more were endangered. From Michele, I learned about the importance of free-flowing rivers. The Colorado, like many rivers in the United States and worldwide, had been dammed, and those dams had changed the character of the river and the composition of its fish community. Once dams go up, they alter a river fundamentally, and it was clear that undoing those changes was not an easy fix.

Worldwide, there have been more than eight thousand introductions of fish species into river basins outside their native range. In fact, freshwater fish are among the most introduced animals in the world. While some species are transplanted on purpose, for sport fishing or to boost fish stocks, many have become established in new habitats after being introduced for aquaculture and escaping their pens and ponds. Tilapias, the common name for close to a hundred species of cichlids from Africa, are an example of the latter scenario. Farmed in eighty-five countries around the world, tilapias have become permanent fixtures in many rivers where they historically don't belong.

During our rafting trips down the Colorado River, we had come across plenty of non-native fish. Sometimes we would go electrofishing, which is a common scientific survey method used to sample fish populations for abundance, density, and species composition. When performed correctly, electrofishing results in no permanent harm to fish, but it is an illegal practice in most of the United States for recreational anglers. Once, as we were working below the Glen Canyon Dam, we shocked up a giant common carp that must have weighed upward of fifty pounds. It was a fish that really should not have been there.

As the fate of the Colorado pikeminnow shows, large fish may be threatened by newcomers that are often small but more adaptable to conditions altered by dams. But large fish do have an advantage over smaller fish: they tend to not get eaten. If you put a catfish into a Nevada stream with small, native fish, those natives will all be gone. Put the same catfish in a river with sturgeon, and the catfish may be the one in trouble. But what if you take a very large fish and put it into a new environment? It might wreak some serious havoc, which is what happened when Africa's largest freshwater fish, the Nile perch, came to East Africa's Lake Victoria.

Also known as the African snook, the Nile perch (*Lates niloticus*) is a percoid (a grouping that includes snappers, groupers, and bass) that is historically found throughout much of sub-Saharan Africa, including several East African lakes. It did not, however, occur in Lake Victoria—the second-largest freshwater body in the world, which is shared among Uganda, Kenya, and Tanzania—much to the chagrin of the British colonizers of those countries. In the 1950s the British very much fancied the idea of fishing in Victoria for the goliath perch rather than the five hundred or so cichlid species native to the lake and which the Brits dismissed as trash fish.

So the same Brits proposed taking perch from Lake Turkana in Kenya and Lake Albert in Uganda and putting them into Lake Victoria. Scientists, and the few others who knew something about the potentially harmful ecological effects of such introductions, thought this was a bad idea. At least, they said, the Nile perch should first be introduced into one of Victoria's satellite lakes as a sort of trial run. It was a compromise the British administrators ostensibly agreed to but apparently had no intention of adhering to; they proceeded to slip an unknown number of juvenile perch into both Victoria and Kyoga, another Ugandan lake.

At first, very little happened; the perch were not even noticed in Lake Victoria until the early 1960s, by which time the British were forced to relinquish their colonial rule and so never got to try their hand at fishing for the new arrivals. Although the perch population seems to have increased at a rapid rate throughout the 1970s, it was not until the early 1980s that, suddenly, the species seemed to be everywhere. Not only had their numbers multiplied spectacularly, the perch had also grown enormous. While the fish in Turkana and Albert, where the perch had first been taken from, reached maximum weights of less than two hundred pounds, the fish in Victoria could grow well over six feet and a monstrous four hundred pounds, or even more.

It was clear what they were bulking up on: the cichlids. As the perch spread, cichlid populations crashed. Later research would show that at least two hundred species of cichlids, many of them popular aquarium fish, were completely wiped out from Lake Victoria in a very short time. "They could not get away from this new monster, which could swallow a hundred fish at once," says Matthew Mwanja, a Ugandan fish geneticist. With female Nile perches able to expel up to twenty million eggs, the species had no problem proliferating, causing enormous ecological change.

At first, the Nile perch was not appreciated as a food fish locally; the fish was oily and had a bad smell. But attitudes toward it soon changed; demand for the fish from abroad skyrocketed, in particular from Europe, where the perch came to be marketed as whitefish. A fishing frenzy soon ensued on Lake Victoria, and by the late 1990s the small fleet of artisanal fishing canoes that had been operating on the lake had expanded to two hundred thousand boats, many of them larger and motorized, scrambling for their share of the catch. The fish became Uganda's second largest export, after coffee, with dozens of fish factories built along the Victorian shores.

However, with time the fishing mania would take its toll. By 2008, the number of perch caught in Lake Victoria had dropped to half of its 1.6-million-ton peak six years earlier. By then, even the tiniest, immature perch were being targeted, in part to feed the black market for the fish in neighboring Congo. Chinese buyers had become well established around the lake, because dried swim bladders from Nile perch sold in China for alternative medicine. As perch numbers decreased, many cichlids actually bounced back. Apparently they had not been completely wiped out after all.

Still, the story of the Nile perch in Lake Victoria, which came to be the subject of an Academy Award-nominated documentary called *Darwin's Nightmare,* has been held up as an ecological horror example of what can happen when non-native fish are introduced into a new environment, and conservationists today often cite invasive species as one of the major threats to the health of freshwater ecosystems worldwide. Excessive predation, as in the case of the Nile perch, is just one of many detrimental impacts from non-native fish being introduced into new habitats; others include hybridization and disease transmission.

At the same time, it's important to point out that most introduced fish—and indeed the majority of all species that take root in new ecosystems around the world, whether by design or default—end up causing little significant environmental damage. Out of twelve thousand alien species of animals and plants recorded in European countries, only 11 percent are deemed invasive. In recent years, a school of thought has even emerged among a minority of scientists positing that the issue of invasive species has been overhyped. Not only do most freshwater fish introductions cause little ecological impact, the argument goes, but non-native species can bring great societal benefits, not the least for recreational and commercial fishing.

Economic benefits can be ephemeral and hard to gauge. On Lake Victoria, the stocking of Nile perch may have created an enormously valuable commercial fishery that at its peak employed hundreds of thousands of people. But it permanently changed the natural system in one of the world's great lakes, and eventually the perch fishery, too, declined and disintegrated into corruption and calamity. In 2017, the Ugandan president, Yoweri Museveni, launched a military campaign to eradicate illegal fishing on the lake, which led to many poor fishermen losing their livelihoods and some allegedly even their lives to violence perpetrated by soldiers.

On the Colorado River, anglers may enjoy being able to fish for bass and trout, but the dark truth of invasives in the Colorado is that the costs, both environmental and economic, dwarf the benefits, and their rise ultimately means we have failed.

While the Colorado defines the southern part of the American West, the Columbia River plays an equally outsized role in the northern region. Comprising parts of seven western states and two Canadian provinces, it is the fourth largest river by volume in the United States, and its basin drains an area the size of France. But like the Colorado, the Columbia has been completely transformed by dams.

The Columbia may in fact be the most heavily dammed major river system in the world, with more than four hundred dams constructed in the basin, mostly built to produce hydroelectricity. Hydropower has been credited for a lot of the economic growth in the region, and still accounts for well over half of electricity generation in the Pacific Northwest. But it's a growth that has come at a huge cost, especially to fish.

For thousands of years the Columbia has been one of the world's great salmon rivers. Going back a century, at least ten million adult salmon and steelhead trout annually returned from the sea to the Columbia River system to spawn. Today, that number has been reduced by 98 percent, according to some estimates. By blocking migration routes for traveling salmon, the dams are the main cause of the fish decline, with eight large dams—four on the Columbia and four on the Snake River, its largest tributary—the worst culprits.

Experts have tried to ameliorate the loss of Columbia's wild fish by installing ladders that allow the fish to get past the dams, and by putting fish in barges and trucks to be transported around dam structures. But despite more than $16 billion having been spent on such efforts over the last several decades, the decline in salmon stocks has not been stemmed. And it's not just the salmon that have been hurting. Their decline has had a serious knock-on effect, since salmon are a critical food source for a wide range of animals, including endangered populations of orcas (killer whales) that feed on Chinook salmon in the sea.

One large fish species in the Columbia River that eats salmon—and one that has also suffered because its migration paths have been blocked by the dams—is the white sturgeon (*Acipenser transmontanus*). The white sturgeon is one of two species of large sturgeon found in the Pacific Northwest—the green sturgeon (*Acipenser medirostris*) being the other. By large, I mean very large. While the green sturgeon can grow seven feet long and up to 350 pounds, the white sturgeon is said to get up to twenty feet long and a monstrous 1,800 pounds. That, of course, would make it by far the largest freshwater fish in the world—except, like most sturgeons, the white sturgeon is anadromous and so not categorized as a true freshwater fish.

Like other sturgeon, it is extremely long lived. There are confirmed reports of individuals over one hundred years old. Very old white sturgeon can still be found in the Columbia River and its tributaries. What you won't find there, however, are many young fish. Not only have the dams blocked spawning grounds for the white sturgeon, they've also disrupted the very sensitive water temperature and substrate conditions that the species requires for laying its eggs. It should come as no surprise that the white sturgeon population that has an unobstructed access to the ocean is also the strongest, while populations become weaker progressively farther upriver, with fewer juveniles.

To see how the white sturgeon might fare in more natural conditions, in the late summer of 2008 I traveled to another river running through the Pacific Northwest, but on the Canadian side: the Fraser River. I was told the white sturgeon population there was thriving, and I was curious to find out why.

Starting in the Rocky Mountains, the Fraser River is the longest river in British Columbia, flowing for about 850 miles through a relatively unspoiled landscape of coniferous forests and floodplains before reaching its outlet near Vancouver. It is the only undammed river holding white sturgeon and the only place where populations of the species have not been affected by fish culture or hatchery activities. Plans to dam the river have met fierce opposition from fisheries and environmental groups seeking to protect migratory fish populations. As a result, the Fraser River has remained undammed and able to maintain the largest salmon population in British Columbia, which in turn has benefited the white sturgeon.

Near the town of Mission, Stefan and I met up with members of the Fraser River Sturgeon Conservation Society, an environmental alliance that had formed a decade earlier amid disturbing signs that the white sturgeon was failing in the Fraser as well. In the mid-1990s, dozens of giant and primarily female fish had inexplicably washed up dead on the banks of the river, while the total sturgeon population dropped below forty thousand, far fewer than the numbers historically found in the river.

Founded by Rick Hansen, a paraplegic athlete and Canadian national hero, the conservation effort received support from many different corners of society, not least aboriginal peoples of the region. Culturally strongly connected to the white sturgeon, the First Nations people, as they are known in Canada, had been instrumental in stopping the harvest fishing of sturgeon in the Fraser.

Recreational anglers also played their part. With strict catch-and-release regulations for sport fishers imposed, a government research program had been set up to encourage angling volunteers to tag sturgeon they caught. A widely successful program with tens of thousands of sturgeon tagged, it yielded invaluable information about their abundance, movements, and growth. A decade into the program, the decline in the white sturgeon had been reversed, with scientific estimates suggesting there were now fifty thousand fish of the species in the river.

On a drizzling Sunday morning, we joined Fred Helmer, a fishing guide who had fished for sturgeon on the Fraser for more than twenty years. He was one of more than two hundred guides working a burgeoning sport-fishing business worth an estimated $20 million a year. Tying up a ball of salmon eggs in panty hose, which he used for bait when fishing for sturgeon, Helmer told us the sport fishery was as good as he had ever seen it. I believed him, as we quickly caught several juvenile sturgeon, an unmistakable sign that the sturgeon population was on the rebound in the Fraser.

As the day drew to a close, and the rain picked up, our largest catch was only a foot long. But then Hellmer got a call. Downstream, someone had made a big catch. We set off in a hurry in Helmer's speedboat. A recreational angler named Stu Love had managed to reel in a big white sturgeon: eight and a half feet long and close to three hundred pounds. It had taken him more than two hours to bring in the beast, which had twice jumped completely out of the water. Helmer and I scanned the calm fish, a female, to see if it had previously been tagged. It hadn't, so we injected a tiny tag behind its skull plate. We took some photos and released the fish back into the turbid waters.

Like all the white sturgeon in the Fraser River, this one would soon head to the sea as part of its natural migration. For sturgeons in the Upper Columbia and other dammed rivers, it is a movement that has been denied them. There are at least eighteen populations of white sturgeon in the Pacific Northwest that have become isolated from the ocean, mostly because of dams.

One such landlocked population is found in the Kootenai River, a Columbia tributary that starts in Canada and runs through Montana and Idaho before returning to Canada. This population is of particular interest because it has been landlocked for many thousands of years and can teach us about how fish are able to adapt.

A long time ago, giant white sturgeon swam freely from the Pacific Ocean up the Columbia River and into the Kootenai, all the way up to Kootenai Falls, near Libby in Montana. The fish spawned in those upper reaches of the river, then traveled back more than seven hundred miles to the Pacific to feed. But around eleven thousand years ago, during the end of the last ice age, retreating glaciers created a waterfall, Bonnington Falls, just upstream of where the Kootenai River flows into the Columbia River, and the fish above the falls could no longer move to and from the ocean.

Unable to reach the sea, the landlocked sturgeons adapted to a life lived solely in fresh water. The isolation had consequences for the growth of the fish, shrinking the Kootenai sturgeon in size. This is not surprising; anadromous fish like sea-run browns and steelhead have access to marine food and resources not available to their freshwater cousins and often grow bigger.

But in time, the Kootenai sturgeon came under serious threat from humans. Overharvesting was a problem, as were contaminants from mining operations in Canada. But what really threatened the future survival of the Kootenai sturgeon was a dam built in Libby in 1974. The dam eliminated nearly all natural spawning conditions for the fish by disrupting natural flows and river temperatures. Almost no reproduction in the sturgeon population has been observed since the dam began operating, and the Kootenai sturgeon are aging and dying. It is estimated that fewer than fifty wild individuals will survive until 2030.

To boost the population of the white sturgeon, the Kootenai Tribe of Idaho built a hatchery on the banks of the Kootenai River where it developed methods to breed the fish in captivity, taking care to preserve as much genetic diversity in the fish as possible. The aim of this program has been to mitigate the loss of natural reproduction by releasing juvenile sturgeon into the river each year, essentially to stave off the extinction of the fish population and afford the tribe and its partners more time to learn how to restore natural reproduction.

The hatchery may stop the Kootenai white sturgeon from going extinct, and that's a good thing. But the problem that keeps wild sturgeon from reproducing in the river remains. A hatchery can buy you time to restore the river, assuming there is the knowledge, money, and political will to do so. But those restoration efforts often fall short, in which case species like the white sturgeon will depend on hatcheries in perpetuity.

What is the point of saving a fish if its habitat is so polluted, so over-run with invasive species, so dammed or overfished that it can no longer support life? Our noble efforts may ensure the fish's survival, but without additional efforts to restore rivers, it may be condemned to an existence in a tank, hatchery, or zoo.

It is said that a river's destiny is to reach the sea, and for millions of years the Colorado River fulfilled that purpose. Then, in the span of a few decades, the river was drained and deprived of much of its life-giving force, until 1988, when, for the first time in its history, it could no longer complete its journey to the gulf. Since then, the Colorado has simply petered out in the desert, becoming the only major river in North America that once reached the ocean but now does not.

It's hard to think of a river as exploited by humans as the Colorado, which makes it surprising to some that it is also considered the birthplace of the modern environmental movement in the United States.

Water in the American West played an important role in the early goings-on of the country's conservation movement. When the Colorado River Compact was signed in 1922 to codify the division of water rights for agriculture and economic development, it included no environmental protections. Similarly, during the dam-building boom on the Colorado and elsewhere, very little attention was given to the potential impact that dams would have on the river's ecology and on fish. But after World War II, things began to change.

In a booming postwar economy, people could afford cars, which they drove to visit natural places. More people began to recognize the costs of environmental negligence, including air and water pollution. In *A Sand*

County Almanac, published in 1949, Aldo Leopold, whom some would call the father of wildlife conservation in the United States, wrote that maintaining the "beauty, integrity, and health of natural systems" is a moral and ethical imperative.

Many people who described themselves as conservationists began to move toward political action, and increasingly their energies became focused on river protection. A particular target of their ire emerged in the 1950s: the Colorado River Storage Project, a proposal by the US Bureau of Reclamation to build a series of new dams in the Colorado River basin. A central feature of this project was going to be a 529-foot-high gravity dam on the Green River, a major tributary of the Colorado, to be built in an area known as Echo Park located inside the Dinosaur National Monument, which spans the border between Utah and Colorado.

The Echo Park Dam would flood a scenic canyon flanked by enormous sandstone cliffs, as well as much of the Green and Yampa River valleys inside the national monument. It is an area as beautiful as the ones where I had spent my summers in college. Conservationists warned of devastating ecological consequences. But the Dinosaur National Monument was a remote part of the national park system that few people had visited, and some wondered why the area was deemed so valuable, especially since the dinosaur fossils that had been excavated there were not in danger of being flooded.

So conservationists set out to build public support for their case, and did so by enlisting the help of influential people like Bernard DeVoto, a conservation writer who wrote an essay in the *Saturday Evening Post* called "Shall We Let Them Ruin Our National Parks?" Soon, coverage of the controversy popped up in major newspapers across the country. Meanwhile, David Brower, executive director of the Sierra Club, made two films about the Dinosaur National Monument and arranged for one of them to be screened frequently in the halls of Congress. At a congressional hearing in early 1954, Brower also boosted the conservationists' case when he showed how the dam engineers had made serious errors in their projections.

Following years of debate, the plans for the Echo Park Dam were eventually scrapped, with legislation signed by President Dwight Eisenhower paving the way for a campaign to establish a national wilderness preservation system. The conservation victory proved to be an important milestone in American environmental history, and many experts date the origins and emergence of a coherent "environmental movement" to the battle against the Echo Park Dam.

As dam building peaked in the 1960s, rivers and fish became a corner-stone of the environmental movement that emerged as a cultural and political force at the time. Of the seventy-five species listed under the Endangered Species Preservation Act of 1966, nineteen were fish, including the Colo-rado pikeminnow. In 1968, Congress established the Wild and Scenic Rivers Act to protect sections of free-flowing rivers that "possess outstandingly remarkable scenic, recreational, geologic, fish and wildlife, historic, cultural, or other similar values."

During the following fifty years, 209 rivers in the United States have been afforded this designation. But not the Colorado. Instead, a tug-of-war over water rights has continued along America's most iconic river, as it dries up further because of drought and a changing climate. Between 2000 and 2014, annual flows in the Colorado River averaged 19 percent below what they were in the century before. At the time of this writing, the water in the basin's artificial lakes has dropped to such precipitously low levels that the federal government has declared a first-ever official water shortage. There is reason to believe this will be the new norm.

And yet I wonder if the Colorado River could one day be brought back to some semblance of its former majesty. After all, we have shown that it is possible, if only temporarily. We did it on March 23, 2014. On that day, the gates at Morelos Dam, situated on the US-Mexico border, were opened to release a "pulse flow" of water into the final, parched stretches of the Colorado River.

The engineered discharge was the culmination of years of negotiations between the United States and Mexico. By taking 105,000 acre-feet of water from Lake Mead and sending it south, we mimicked, on a small scale, the spring floods that historically inundated the Colorado River delta, in the hopes of restoring some of the natural bounty once found there. As the water made its way across the salt-crusted sand plains, the place that Aldo Leopold after a canoe trip in 1922 called a "milk and honey wilderness" of "a hundred green lagoons" once again sprang to life.

Children who had never seen the river bathed in it, as the skies above filled with hawks, egrets, and ospreys. Then, nearly eight weeks after the release, the river finally, and briefly, reunited with the sea, and in that fleeting moment, the Colorado River was once again complete. If you closed your eyes, you could envision the fish returning—the humpback chub, the bonytail, and the largest pikeminnows that anyone has ever seen.

Monsters?

During a morning run on September 19, 1996, two SEAL instructors from the US Naval Amphibious Base on Coronado Island just off San Diego, California, made a strange discovery: a freshly beached silvery fish more than twenty feet long, whose head and tail had been detached. An examination by scientists from the Scripps Institution of Oceanography in nearby La Jolla soon concluded that the animal was an oarfish, a large, elongated lampriform that is found in most ocean waters yet rarely seen.

Word of the discovery made its way through the naval base, and a photograph of the mammoth snakelike fish, held up by at least a dozen members of the base's first-year class of SEALs, was taken by a Scripps photographer named Leo W. Smith. The photograph appeared in many newspapers and magazines throughout the United States and Europe.

Then something strange happened. The same photograph began to appear on postcards widely circulated in Southeast Asia, many of which were pinned to the walls of bars and restaurants in Thailand and neighboring countries. Only this time the photograph carried a caption that read: "Payanak, Queen of the Nagas, seized by American Army at Mekong River, Laos Military Base on June 27, 1973, with the length of 7.80 meters."

Also known as Phaya Naga, the payanak is a mythical animal with a single, dragonlike head and long body that features prominently in Southeast Asian folklore and Buddhist belief. Represented on temples throughout Thailand, Cambodia, and Laos, it is said to be the lord of nagas, serpentlike deities that live in the Mekong River, in the border region between the natural realm and the netherworld, where they protect treasures by spitting fireballs at those who dare to come too close.

The story of how the oarfish became a naga—or how a marine fish that washed up on a beach in California could give shape to a fictional river monster said to live on the other side of the world—encapsulates how easy it is to blend fact and fiction to create a fake fish story and pass it off as true. I have come across countless photos, illustrations, and drawings of unreasonably large fish doing the most absurd things (like breaking through a concrete dam). Those are easy to dismiss. It's trickier when the giant fish shown in photos actually exist but are made out to be something they're not. The photo of the naga was believable because the giant oarfish is a real fish, not a fireball-spitting serpent of the netherworld.

The story speaks to our general fascination with aquatic monsters; the main character in the book of Jonah is swallowed by a "great fish," and Leviathan in the book of Job is described as a primordial sea serpent. These creatures begin as myths, then acquire a corporeal reality as their true existence becomes known. Thus the mermaid metamorphosed into the manatee; the popyp became the octopus; and the kraken, the cephalopod-like sea monster that terrorized seafarers in the ancient Norse sagas, is now believed to be a giant squid.

While the majority of mythical water monsters appear to prefer the seas, many a monster tale has also arisen out of rivers and lakes, the most famous being Nessie, the Loch Ness Monster, in Scotland. We find many lake monsters in the United States too, including the Bear Lake Monster, a serpentlike creature said to maraud the shoreline of a lake on the Utah–Idaho border; the Iliamna Lake Monster, also known as Gonakadet to the inhabitants of the small Alaskan fishing village of Iliamna, who claim to have seen a beast with a wolf head and an orca body; and Tahoe Tessie, the dinosaur-like being residing in North America's largest alpine lake, Lake Tahoe, not far from Reno, Nevada, where I live.

The monster in all its forms has been a staple in American entertainment culture. A keyword search of "monster" on IMDb, the online database of information related to films and television programs, generates more than five thousand titles. Two of those refer to our TV show, which in North America was initially called *Hooked: Monster Fish* before being changed to *Monster Fish.*

At first, I hated the title, hooked or not. "Monster" suggested an imaginary creature capable of doing great harm to its unsuspecting victims, but that's not the way I thought about these animals. Not only were the fish real, they were rare and in many cases critically endangered, in dire need of conservation. For the most part they were not dangerous, and I was uncomfortable with the idea of reinforcing that stereotype.

Soon after the National Geographic Channel greenlit the show, I attended Nat Geo's annual Explorers Symposium. I met with folks from both the channel and the society, the nonprofit scientific and educational arm of National Geographic, which was providing financial support for my Megafishes Project. I raised my concerns about the show's title.

I was particularly irked by the use of the word "hooked" in the title. The promotional poster showed a close-up of a giant fish about to get hooked, and there was even a widget game based on the show in which the aim was to hook a fish. The image missed the point. My connection with these giant fish was not as a fisher, but as a scientist and conservation biologist studying species on the edge of extinction. The idea of fishing for them—or making a game about fishing for them—was not where my head was.

The National Geographic people, especially those with the society, were sympathetic to my concerns. But there were arguments in favor of the "monster" title. It was the title that had been used to launch the show, and so it had a built-in recognition factor that would help us reach the largest possible audience. I well understood the potential of a television show—especially one distributed by a top media house like National Geographic—to communicate an important message about endangered species and to inspire people to protect the environment. This outweighed my concerns about potentially sensationalizing the name of the show. Plus, "monster fish" is a common term used by fishers and anglers to describe a particularly large fish. Seen in that light, it seemed like a good title for a TV show about the world's largest freshwater fishes. In any case, I didn't want to push things too hard; I was new to the business of making television shows.

What I did not want to compromise on, however, was the educational and scientific aspects of the show. I was sensitive to the fact that *Monster Fish* might be the only opportunity for some viewers to learn about these animals, so I wanted to make sure that we presented the information in the show as accurately as possible. The easiest way to do that was to engage with scientists and local fishers and allow them to speak for themselves, about the places where they lived, worked, and fished, and about the animals they caught or studied.

We would fish for species only when recreational fishing was encouraged or could provide us with data for research. The basic storylines of the show were scripted. The script provided guidance on whom we would meet and the locations we would visit, and each episode tended to follow a structure that built toward catching a big fish. With less than two weeks in the field for each shoot, we split our time between finding the biggest fish and gaining an appreciation for the ecology and conservation status of a particular species.

Some shows included more science than others, but it was important to always get right whatever science a show did feature. As a biologist myself, I was cognizant of how annoying it was to scientists when subjects were oversimplified or exaggerated. I tried to avoid reinforcing erroneous preconceptions or stereotypes, and accuracy was particularly important when talking about animals that few people knew much about.

It was motivating to see the reaction that people had when they learned about giant fish. It was clear that many people didn't have any idea that these creatures existed in our rivers and lakes. Everyone had heard of the Loch Ness Monster, but, for the most part, people didn't know about the colossal catfishes, giant stingrays, and awesome arapaima that were actually real.

It made me think of the story of the oarfish and the naga. The claim that the silvery fish depicted in the photo was actually a mythical Mekong River monster had long been debunked. Yet it seemed like all the restaurant and bar owners had kept the photo up on their walls anyway. It could be that they were not aware that the photo was a hoax. But maybe there was another explanation. Maybe the monster seemed just as believable as the real-life giant.

We won't dwell much further on imaginary beasts when the real-life creatures are so much more fascinating, but it's worth examining the story of the Loch Ness Monster for just a moment, because there could be a compelling connection between Scotland's famous lake lodger and the largest freshwater fish in Europe: the wels catfish.

No aquatic monster has received as much attention as Nessie, whose legend can be traced back to AD 565 when the Irish missionary Saint Columba is said to have encountered a giant beast in the River Ness in the Scottish Highlands. The first modern sightings of a monster in Loch Ness were made in the 1930s, when the *Inverness Courier* reported on a whalelike creature sending the lake's water "cascading and churning." A respected British surgeon, Colonel Robert Wilson, claimed he took what would become a famous photograph of the long-necked monster while driving along the lake's northern shore.

Since then, a hodgepodge of scientists, naturalists, and amateur sleuths have descended on the second-deepest lake in Great Britain to solve the mystery, an effort that has continued unabated even after Colonel Wilson's photograph was revealed sixty years later to be a hoax that used a sea monster model attached to a toy submarine.

The most dedicated hunter of the legendary monster must be Steve Feltham, who in 1991 sold his house in southwest England, left his girlfriend and a profitable family business installing burglar alarms, to move to Loch Ness to search full time for proof of Nessie's identity. For almost a quarter of a century, Feltham monitored the loch from a lakeside hut lacking both water and electricity for any signs of the monster, until, finally, in 2015, he concluded that Nessie was in all likelihood not a prehistoric monster but a very large wels catfish.

The wels catfish (*Silurus glanis*) is truly a real-life monster. Also known as the European catfish, or sheatfish, it is the largest-bodied freshwater fish of Europe, potentially growing more than sixteen feet long and possibly attaining a weight of more than 300 kilos (660 pounds), according to Fish-Base. That, of course, would put it in contention for the title of world's largest freshwater fish.

But it's not just its massive size that makes the wels a monster. With its mouth lined with rows of small teeth and two long barbels on its upper jaw, it has a fearsome appearance. Scientists have shown that the wels can track prey in three dimensions in absolute darkness using its whiskers and lateral line, the visible line along the side of a fish that consists of a series of sense organs to detect pressure and vibration. As a gluttonous carnivore, it preys on a wide range of animals, including frogs, mice, rats, ducks, worms, snails, insects, crustaceans, and most fish it comes across. There are even strange stories going back to the Middle Ages of wels catfish flinging themselves onto land to snatch babies and farm animals.

While that may seem farfetched, I did find video clips of the wels on YouTube that captured my attention. They were from the town of Albi, in southern France, and showed catfish in a river jumping out of the water to catch pigeons on the shoreline. The scenes were reminiscent of orcas beaching themselves to feed on seals, but I had never seen anything like it in a freshwater fish. It was a visually striking behavior that would make for good television, and we quickly put together a pitch for a show on the wels catfish, which National Geographic enthusiastically approved.

One of eighteen *Silurus* species, the wels catfish is native to east and central Europe, basically the area east of the Rhine River. It's a valuable food fish in countries like Hungary, Poland, and Slovakia, where it's both caught in the wild and farmed and exploited for its tender white meat, as well as its skin (for leather and glue production) and eggs (for caviar). But the wels has also been introduced into several western European countries for recreational fishing, including the Tarn, which is a tributary of the Garonne

River, running through the small, historical French town of Albi, where the YouTube videos had been filmed.

The pigeon attacks had first been observed a few years earlier. The best place to catch the spectacle was from atop Pont Vieux, an eleventh-century bridge that runs across the Tarn and connects the two separated parts of the town. Below, there were two islands, one on either side of the bridge, on which scores of pigeons wandered about. From that vantage point, we could clearly see the contours of several huge catfish that slowly circled the islands, looking like sharks on the prowl.

Not long after we had set up to film from the bridge, we witnessed our first attack. As one solitary pigeon strayed into the water to drink, we could see a big fish edge closer and closer to the shore, until it was right next to its unsuspecting prey. I wanted to yell out to warn the poor bird, but resisted the urge. Suddenly the catfish catapulted out of the water, snatching at the pigeon. A flurry of feathers followed, and I was sure that the fish had got a hold of the bird. But somehow the pigeon escaped, flapping to safety while the catfish slipped back into the river.

One would think such an encounter would send a clear message to all the pigeons that walking around an island stalked by murderous catfish was not the greatest idea. But no: after initially moving away, the pigeons soon returned, even the one that had narrowly escaped death. This would go on for hours. Birds would edge into the water, moving in twos and threes. Likely attracted by the splashing and the oils coming off the birds' feathers, the giant fish would come closer and closer, and then suddenly leap out of the water in a spectacular burst of energy. The fish failed to catch the birds, but it was still shocking to watch.

Usually, the catfish that attacked the birds were of medium size, three to five feet long. The local biologists had a theory about that: the smallest wels, they speculated, weren't big enough to eat pigeons, while the largest fish were too big to propel themselves out of the water. But while we were there, the medium-sized fish didn't succeed either. All day they swam back and forth along the bank, waiting for the right opportunities to attack. When they did, they repeatedly failed. Usually, the attempts were clumsy, resulting in a lot of splashing but no pigeon-eating.

The scientists helping us, Julien Cucherousset and Frédéric Santoul, theorized that the degree of the slope of the gravel bank had recently changed, and that that may have accounted for why the fish weren't able to get to the pigeons. In truth, I didn't mind. The footage was spectacular, and I got to witness the whole predation sequence up close. I didn't need to actually see the catfish eat the pigeon. It was clear that these attacks actually happened, that it wasn't a made-up story.

Although the wels catfish I saw in Albi were large, they were not as enormous as the giant catfishes in the Mekong. They would continue to grow, however, and maybe there was reason to believe that, in the future, they too could reach record sizes. After all, the wels catfish, like many megafish, can live for a long time, up to eighty years, and unlike most giants, they were thriving not only in the Tarn but in other rivers around Europe. Only a few months earlier, an angler had hauled in a 280-pound catfish from the Po River in Italy. There were old, unconfirmed reports of the wels growing larger than the Mekong giant catfish, and as I left France I was not ready to count it out as the megafish champion.

The wels catfish may not, however, be the monster in Loch Ness. A few years after Feltham, the Nessie hunter, had drawn his conclusion, a team of researchers led by Professor Neil Gemmell, of the University of Otago in New Zealand, conducted a new study, which suggested that the legendary sightings of Nessie could be due to big eels in the lake. Test samples, the researchers said, had shown large amounts of eel DNA and none from catfish. But, Gemmel added, "for those who still want to believe in monsters, there is still a lot of uncertainty in our work," adding that "the absence of evidence isn't necessarily evidence of absence."

With *Monster Fish* in its third year, I wanted to get back to the Amazon, where we had shot the first National Geographic production. That show had centered on the arapaima, but the main impression the Amazon had left on me was the vastness of the river system; how unpopulated it seemed in comparison to rivers like the Mekong. I wanted to do a show from the Amazon that conveyed the wildness and mystery of the place—and to follow in the footsteps of an adventurous US president.

In 1913, Teddy Roosevelt led a legendary expedition to navigate the until-then unmapped Amazonian river called Rio da Dúvida, or the "River of Doubt." I had read Roosevelt's *Through the Brazilian Wilderness,* his account of a treacherous journey that had resulted in the deaths of several people in his party and left a disease-stricken Teddy fighting for his own life deep in the Brazilian jungle. A best-selling book by Candice Millard called *River of Doubt* vividly described the former president's torturous trek down the rapids-choked river, which was later named the Rio Roosevelt in Teddy's honor.

What had particularly captivated me was Roosevelt's accounts of huge fish in the wild and winding waterway. The former president wrote of massive catfish, nine feet long, with "monkeys in their stomachs," more dangerous than the caiman that patrolled the river. I was skeptical the fish were truly dangerous—Roosevelt claimed they even ate local children—but I believed his traveling party must have encountered some fearsome goliath catfish.

Those few fish accounts aside, Roosevelt's expedition had catalogued mainly terrestrial species, not aquatic life. No one had, to my knowledge, focused specifically on the river's freshwater life. So I thought, "Why not try?" Not necessarily as a comprehensive biodiversity survey—that wouldn't be possible to do in two weeks—but an attempt to see if the giant fish that Roosevelt had come across on his pioneering journey were still there.

It was going to be a challenging adventure. Originating in the Rondônia highlands, the Rio Roosevelt runs through some of the most isolated jungle on Earth. Navigating it is made particularly complicated by the many class 3 and 4 rapids that propel the river forward on its 470-mile journey to the Rio Aripuanã, a tributary of the Madeira. It was a chance to do something very few people had done before, and despite the uncertainties, I felt like we had a plan that would work well. Looking back, I shake my head at the difficulties we encountered—even before our boats were on the river.

Our National Geographic film team met up in Manaus, where we revisited the colorful fish market before we flew southwest, across the Amazon, to the city of Porto Velho. From there, we drove to a location far up the Rio Roosevelt watershed, the starting point of our descent down the river. It had taken Roosevelt's expedition a month just to get to the river, but for us it was a day's drive, through forest land that had been cleared for ranching, past still-smoldering fields, onto smaller and smaller dirt roads that eventually dead-ended at a ranch house on the edge of the jungle. Through the trees, a river, set deep and flowing fast in a stone channel no wider than ten feet opened up. We had found the River of Doubt.

By then, our team had expanded to include two Brazilian biologists, a Brazilian fishing guide, and three members of an indigenous group, the Cinta Larga, who we hoped could liaise with local communities and, if need be, negotiate our passage. We set up camp in the grass in front of the ranch house and were planning to hit the river the next morning. The goal was to travel down as far as possible and exit at one of the few access points between the ranch house and the confluence of the Rio Roosevelt and the Rio Aripuanã, hundreds of miles downstream.

That evening, however, I became violently ill. I lay in my tent, unable
to help as the team unpacked and repacked supplies, inflated rafts, checked
motors and gear. I knew the Amazon could be an unforgiving place. In the
wetness and heat of the jungle, sickness lingers and worsens, and infections
fester and grow. Once we were on the river, we might be three or four days
away from a hospital. What if I got worse? It was a big risk, but I didn't want
to throw in the towel before we had even got going.

Thankfully, I felt much better the next morning. Given the speed of
recovery, I didn't think I had contracted anything serious, like malaria, but
just a simple case of food poisoning or heat stroke. I was relieved as we set
off down the river, the group squeezed into the two heavily loaded rafts,
while the kayakers on the team took turns scouting ahead.

The first day was peaceful and uneventful, as we traveled the narrow
sliver of water through nearly impenetrable foliage. As in other remote
places in the Amazon, abundant wildlife surrounded us, and we could hear
the calls and cries of animals in the trees, but we couldn't actually see any
of them, except for birds, which included brightly colored parrots who kept
flying around and calling overhead.

Black piranha (*Pygocentrus piraya*), the largest and most aggressive
piranha species, were everywhere. The pattern was always the same: cast
a live or dead bait, wait five seconds, feel the staccato jabbing and pulling
as the piranha attacked the bait, and reel it in (or reel in a broken line or
bent hook). I suspected we wouldn't see large fish right away, because the
river was still small and there were rapids down the river that would act as
barriers to fish moving upstream.

The troubles started on the second day. The outboard motors for our
heavy rafts broke down, and our guide (and the owner of the motors) was
unable to fix them. Meanwhile, Trip, the kayaker, began feeling sick and
quickly developed a strong fever. It seemed to be something serious, like
malaria, so we created a small area for him to lie down in the back of the raft
before continuing on.

Over the next few days the problems kept piling up. Many of them
seemed to be of the same nature that Roosevelt's team had experienced a
century earlier: seriously ill crew members, uncertainty about our location,
and slow progress. At times, we had to chop our way through fallen trees
that blocked the river. There were rapids and waterfalls that we could not
pass, and several times we had to unpack all our gear and carry it around
the obstacles. It was hot and wet, and impossible for Ryan, the cameraman,
to keep all his equipment—multiple cameras, batteries, lights, microphones,

laptops and external hard drives—dry, and he was obviously struggling. The only positive was that Trip recovered after a few days.

All the while, we had to make sure we respected the sovereignty of the Cinta Larga, the indigenous group whose land we were crossing. The Cinta Larga had remained in almost constant conflict with the outside world ever since they were first contacted a half century ago. As outsiders kept encroaching on their land, burning forest and converting land into pasture, the Cinta Larga fought back. Recently, tensions had escalated after some miners had been killed in the area. It was a volatile situation, and one that required us to follow the judgment of our local guides.

We also had an issue that, while hardly a matter of life or death, was problematic for our mission: besides the piranha, we couldn't catch any fish. The Brazilian fishing guide who was with us was actually from Porto Velho and had never fished the Rio Roosevelt. Aside from him, the fishing team consisted mainly of two Brazilian biologists, whose dip nets and seines were only good for catching small fish, and our local guides, who knew the area well but had never encountered a goliath catfish.

After a week on the river, we'd gone only a third of the way, not knowing what lay ahead. We no longer had enough food for the trip, and our motors weren't working. But there was no going back. We had portaged over small waterfalls, and it would be impossible to return upstream. We had passed the point of no return almost as soon as we started the trip. On a river like the Roosevelt, or even the Grand Canyon, once you start down the river, you are committed. There are often few or no options to get off, so you have to push on. Even so, we recognized we were outmatched: sick, slow, with failing gear, and apprehension starting to rise within the group. It didn't make much sense to continue traveling into the unknown. Even if the situation was not as dire as it had been for Roosevelt's expedition, at least not yet, I could still appreciate the dilemmas those people had faced.

Still, I wanted to get down the river far enough to find the goliath catfish. The kayakers were in it for the adventure and wanted to attempt to complete our journey. The Brazilian biologists, however, were more cautious. One of them, Flavio Lima, argued that the river should not be taken lightly. Yes, we were on an adventure, he said, but it was one that had not gone according to plan, and things were likely to get worse if we continued. He was right. In this environment, imperfections are often magnified, and even small mistakes or accidents can spiral into a major incident.

The discussion came to a head one night as we camped near another set of rapids, which we had not seen on the satellite maps and that we would have to descend the following morning if we continued. As I settled into my small tent to sleep, I could hear the American kayakers and the Brazilian biologists sitting around the campfire in an animated discussion. Apparently we were nearing our last exit point if we decided to call it quits. The Americans wanted to continue on, but the Brazilians warned that each rapid was a reminder of the obstacles we'd likely face if we did so: portages, breakdowns, and possible injuries.

As I drifted off to sleep, Flavio talked about bullet ants, which live in the Amazon rainforest and are considered to have the most potent and painful bite in the whole insect kingdom. When I woke up in the morning, a decision had been made to evacuate from the river. Although I had wanted to go on, I was also happy we didn't, because it was clear there would have been no turning back, regardless of what befell us or how long it took. It was the right call.

It may sound like our trip was a disaster, but I didn't view it that way. We visited a place that few people will ever see. It had been a search for answers—what's out there? what will we find? what obstacles will we face?—much like my quest to find the largest fish is a search for answers.

But it's not always about the answers. It is as much about the question and going out there to try to answer it. This whole project, the search for the largest freshwater fish, was based on a simple question, and it was that question that had taken me on this complex journey, connecting me with people and places. On the River of Doubt, the team had bonded in a way that we wouldn't have off the river. I had not known Flavio before the trip, but I came to appreciate his sense of humor and insight. He and I used his seine net to collect small fish in spots along the river, and later he wrote to me to say that one of the fish we caught was likely a species that had not yet been described by science.

After the trip's premature end, we returned to Porto Velho, near a famous location that I had long read about: the Teotônio rapids, the site of a traditional fishery for big catfish on the Madeira River. So we went there and filmed over several days, watching local anglers fish above the rapids on rickety platforms. It was a spectacular scene, similar to the traditional fishery on the Mekong River at the Khone Falls.

At Teotônio, I got to see the dourada, the gilded catfish, and it brought home to me that often we look for one thing and find another. For me, curiosity drives everything. It is the way to explore, to learn, and to share. We may think we know where we're headed, but we don't really know. We get out there and let life wash over us, like the violent rapids of Rio Roosevelt washed over me.

Like many of the giant fish in this book, animals deemed "monsters" are often simply misunderstood—or turn out to be a myth. Call an animal a man-eater, however, and it's different. A man-eater has deadly intentions; it's right there in the name. It is an animal that will eat you. But are there actually animals that will stalk, kill, and eat humans?

Let's exclude the scavenging of corpses; a single attack born of opportunity or desperate hunger—as in the case of the dozens of starving brown bears that in 2008 killed two geologists working at a salmon hatchery in Kamchatka, Russia—and the incidental eating of a human that the animal has killed in self-defense. We will also ignore the kind of parasitic predation that the biggest animal killer of people by far—the mosquito—engages in. Its blood-sucking nature aside, it's the diseases that it carries that kill. Rather, which animals will stalk human prey and have the propensity for killing and eating humans, not just sucking their blood and spreading disease?

Tigers, for example, are said to be responsible for more human deaths through direct contact than any other wild mammal, with about fifty people killed by tigers in India each year. Experts estimate that out of the three thousand tigers in India today, only a dozen or so may be persistent predators of humans. The majority of people killed are reportedly attacked at random inside the tiger's territory rather than hunted for meat. It is a similar situation with lions and leopards.

Bear attacks are rare, though cases have been documented of rogue brown bears, stricken with disease or hunger, setting upon humans, and young and undernourished polar bears are known to hunt people for food, even if actual killings are rare. Other land animals—such as wolves, hyenas, and Komodo dragons—may kill and eat humans, but not as any kind of normal hunting behavior.

In the water, people may think of sharks as man-eaters, but only a handful of people are killed by sharks each year, and instances of sharks devouring their victims are basically unheard of. Crocodiles are a much more menacing threat to humans, especially the saltwater kind I had seen in Australia and the Nile crocodile found in Africa. Nile crocodiles are estimated to kill—and eat—between 150 and 500 people each year, making them the number one cause of human deaths involving wildlife on the continent.

But are there any freshwater fish that have adopted what could be described as man-eating behavior? While stories of piranhas as people-killers had been greatly exaggerated, there was one freshwater species—and a giant catfish at that—that actually had the man-eater moniker incorporated in its common name: *Bagarius yarrelli,* the man-eating goonch.

Very little is known about the goonch, even by catfish standards. We know it's a big fish, but just how large the goonch can grow is far from settled. While there are credible reports of individuals caught in Indian rivers weighing three hundred pounds, more questionable sources speak of goonch weighing more than six hundred pounds having been captured in Bangladesh. It's fished for food throughout its range, but almost nothing has been scientifically documented about its behavior, feeding patterns, or reproduction, though there seems to be consensus around one fact: the goonch, with its high-set, circular eyes, sharp teeth, and oversized appendages that awkwardly extend from its face, is a pretty gnarly looking creature.

But it's not its looks that has earned the goonch its notorious reputation as a man-eater. This instead stems from its suggested involvement in a series of reported fatal attacks on humans in three villages on the banks of the Kali River in India and Nepal between 1998 and 2007. In one attack, a witness said an eighteen-year-old Nepalese man had been dragged under water by something described as a "mud-colored water pig." It did not take long before people began pointing fingers at the goonch, which is prevalent in the area, as the likely culprit.

But was the goonch really a man-eater, and if so, what had caused it to become one? One theory that quickly gained traction in newspaper reports suggested that the catfishes had developed a taste for human flesh after feasting on half-burnt human remains discarded from traditional funeral pyres held on the riverbanks. Maybe it had expanded its diet to include living people as well? Others spoke of rogue or even mutant fishes with indiscriminate hunting proclivities.

I was skeptical. It's rare for animals, in the wild and unmolested, to be aggressive toward humans. Even sharks, perceived to be among the most dangerous aquatic predators, rarely bite unless provoked. That didn't mean I wasn't curious to investigate the matter of the goonch deaths. Besides, *Bagarius yarrelli* was on my megafish list, and I wanted to know more about it. So I was happy when National Geographic greenlit an episode on the goonch.

In mid-April we traveled to northern India for two weeks of filming.

The crew and I were joined by Vinay Badola, a bearded and slender Indian fishing guide who wore a bucket hat like Gilligan in the 1960s TV show *Gilligan's Island* and who would spend the entire expedition talking nonstop about anything that had to do with the goonch. Vinay loved the goonch.

Our trip began in Jim Corbett National Park, a tiger reserve and India's oldest national park, near the border with Nepal. From there, we visited three rivers: Ramganga, a tributary of the Ganges where a 287-pound goonch was said to have been caught; the Sarayu; and the Kali, where the three fatal attacks had reportedly occurred. As we set off in our four-wheel-drive trucks, a challenging road lay ahead. The landscape was mountainous, with dirt roads up and down steep canyons and along precipitous cliffs. We spent long, hot days in cramped and uncomfortable conditions.

Despite heat, dust, and a thick haze from small cooking fires and the burning of crop waste after harvest, India's rivers ran surprisingly clear, which would help us in our pursuit to film the goonch. But, as Vinay reminded me, the goonch inhabit large river channels typically associated with fast-flowing, turbulent rapids where they take shelter among boulders and large rocks. While known to move in large groups, they can still be difficult to find. Stories told of accomplished anglers waiting months without so much as a bite from a goonch.

Diving in a Hindu temple pool on the Ramganga River brought me face to fin with hundreds of golden mahseer (*Tor putitora*), a beautifully golden-colored freshwater fish that can grow almost eight feet long. The fish that surrounded me were nowhere near that big, most just two or three feet. But it was mesmerizing to watch them circle me, gliding effortlessly through the water, and turning quickly if they got too close, sending off a flash of gold. However, despite a recent photograph of a seven-foot goonch caught nearby, we had no luck finding the monster we sought.

A few days later, at the confluence of the Kali and Saryu Rivers, a procession of several dozen people descended a mountain and headed toward the water. Some were carrying a large bundle wrapped in white cloth, while others hauled large pieces of wood. As soon as they reached the river's edge, they began to build a wooden structure in a shallow portion of the river. Gradually, it grew to a towering platform.

By then, I had realized what was happening. The white bundle was a body and the wooden tower a funeral pyre. Mourners placed the bundle on top of the structure and set it on fire. It burned slowly at first, but then the flames eventually reached ten feet in the air, as the white bundle incinerated.

The burnt wood and ashes crumbled into the river, where they were washed away. The following day there would be no evidence of what had happened, except for a few black rocks.

I knew of "sky burials" in Mongolia where bodies were left to the elements and scavenging animals, and there were certainly scavenging fish, the goonch among them. In the Mekong, the "dog-eating" catfish were said to have grown used to eating the carcasses of all kinds of animals. What was to say that their catfish cousins in India had not done the same, only with humans?

On the Ramganga River, Vinay and I were fishing a shallow riffle above a pool when we got our first bite. Using dead fish as bait, we had cast out into shallow water when suddenly a dark shape streaked across my frame of view, splashing and thrashing around our bait as it came in. Almost as soon as it registered as a goonch, Vinay hooked it, and the fish quickly turned back downstream, pulling the line (and Vinay) toward the deep pool below.

Goonch are strong fish, and they have the ability to flatten their body and wedge themselves between rocks on the bottom. Luckily, this fish didn't get into the rocks, and eventually Vinay was able to reel it in. It was small for a goonch, less than three feet long, but still an impressive fish, with fleshy barbels, multicolored, leathery skin, beady eyes, and nail-like teeth pointed back into the mouth to make it harder for prey to escape.

After releasing the fish, we walked farther upstream to a long, deep pool with a rock ledge on one side and a pebble beach on the other. After fishing for a few hours, with no luck, it was time to get in the water. We had brought snorkels and masks, as well as underwater cameras. After wading into the water, mask on, I pushed off and submerged entirely.

In water too deep to stand in, I could see only about ten feet ahead, and I could barely make out the river bottom, but when I looked closer I realized it was covered with white bones. Then a strange, unnerving sound made its way through the water. Every few seconds, it sounded like someone was running his hand through a pile of shells. It was faint, but I had the unmistakable feeling I was not alone in the pool. Something was moving through the bones, clinking and clicking. Then it sped past me and disappeared. Then it happened again. This time, I got a good look at what it was: a goonch. And not just one. There were many resting at the bottom, on top of the bones. The clicking sounds were the goonch moving over the bones. As I dove down, the goonch spooked and darted away, scattering the bones with their tails as they disappeared into darker water.

My reaction might be surprising, but any fear I had had quickly dissipated, replaced by excitement and the recognition that I was very fortunate to be able to observe these incredible fish so close. And these goonch, which measured between three and five feet, were not shy. Even as I got relatively close to them, they stayed motionless on the river bottom. But if I got too close, they'd flee to a rocky crevasse where I couldn't reach them.

The whole experience mirrored my interactions with other megafish. After an initial "getting to know you" period, I felt comfortable around the goonch, and their behavior became predictable. In fact, goonch turned out to be one of the best fish to swim with and observe because they weren't timid, and there were occasions over the next few days when they would let me approach within a few inches. At one time, a goonch dropped slowly down a submerged rock face with me, to a depth of about fifteen feet, before I had to return to the surface to catch my breath. When I dove back down again, the goonch was sitting just where I had left it, as if waiting for me and as if swimming with me was the most natural thing in the world.

Were they man-eaters? Not in the way that some had described them—of that I was sure. But what about the stories of attacks? The truth is, I don't know. What I do know is that many more people drown every year than are killed by aquatic animals, yet we have a tendency to look for outrageous explanations even when there are simple ones staring us in the face. I know that humans are a much bigger threat to goonch than goonch are to humans. Like with most species of megafish, goonch populations are declining, and the really big goonch, which are the basis of fantastical stories, may already be gone. It is the same story over and over: people killing animals, not the other way around.

Ecosystems are largely defined by their food webs, which is why ecologists are fixated on the question of who eats what in nature. Organisms are classified according to their trophic level, or position they occupy in the food chain, and there are five such levels. Chances are you've seen them depicted in the form of a pyramid illustrating the amount of energy that moves from one feeding level to the next in the food chain. At the first level are plants and algae that make their own food, called producers. The second level includes herbivores (called consumers) that eat those plants. Moving up the pyramid, to the third level, we find carnivores that eat herbivores. And at a level above that are carnivores that eat other carnivores.

Which brings us to the fifth and highest level. This is where the apex (or alpha) predators reside, those carnivores that, when fully grown at least, have no other creatures naturally preying upon them. They are emperors of their own ecosystems, and include some of the most iconic animals on Earth—lions, tigers, wolves, polar bears, killer whales, saltwater crocodiles, and many sharks—but also large freshwater fish, several of which have been mentioned in this book already, like the wels catfish, piraiba, alligator gar, and arapaima. In fact, the first vertebrate apex predator may have been an ancient relative of the lungfish, a three-foot-long creature called *Megamastax* that lived in the Silurian period.

Ecologists pay particular attention to apex predators, and for good reason. These animals have profound effects on the ecosystems in which they live by keeping their prey populations in check and maintaining biodiversity. In some cases, they can even strengthen populations of species they prey on, which is what happened when gray wolves were reintroduced to Yellowstone National Park in 1995. Many had feared that the return of the wolves, which had been killed off in Yellowstone in the 1930s, would wipe out local elk populations. Instead, wolves mainly hunted weak and sick elk, which helped to create more resilient elk herds.

The problem is that, with their own populations largely restricted by the availability of food, apex predators generally exist at low numbers, which makes them vulnerable to being hunted. By whom? you may ask, since we're talking about a class of animals that has no predators. While it's true that apex predators have no natural predators, there is still one species that is more than capable of killing them, and that, of course, is humans.

It's easy to think of humans as an apex predator, since there are few other species out to eat us, but that would not really be correct. Scientists showed this in 2013 when, for the first time, they calculated the trophic level of humans and found it to be 2.2, similar to pigs and anchovies. While meat consumption has been steadily rising in recent decades, most people still eat a primarily vegetarian diet, especially in Asia and Africa, and the meat we do eat comes mainly from herbivores, like cows and chicken, not carnivores. Big meat-eaters are generally killed for trophies or because they're a nuisance rather than for consumption. This is the case with many land animals like lions or wolves, but, as we've learned, it also applies to an aquatic predator like the alligator gar.

Maybe a more appropriate term for humans would be "super predator," which is what a study in *Science* in 2015 (titled "The Unique Ecology of Human Predators") dubbed us. It showed that the human species is unlike any other predator on the planet when it comes to our choice of prey. While predators across the animal world focus their efforts on juveniles, simply because they're easier to kill, humans are more likely to kill big, brawny adults.

And not just a little more likely. On land, the study found, our harvest of adult carnivores, often for trophy hunting, was nine times that of other large carnivores that mostly killed each other through competition. In the oceans the differences were even greater. There, results showed, humans on average kill fourteen times more adult prey than other marine predators, and eighty times or more in extreme cases. While the study did not look at freshwater fish, it's not hard to imagine that the numbers would be at least as high for fish in rivers and lakes. It's easy to see why this is a problem. By focusing on catching large adults, we remove individuals in their reproductive prime that are needed to replenish diminishing populations.

Only relatively recently have we really begun to comprehend the impacts of apex predator loss on ecosystem function, and the emerging picture is grim. Studies have shown that removing apex predators from an ecosystem triggers an ecological phenomenon known as "trophic cascade," a chain of effects moving down through lower levels of the food chain. But the impacts can also be surprising and far-reaching, including changes in vegetation, wildfire frequency, infectious diseases, carbon sequestration, and water quality. As one 2011 study in *Science* put it: "The loss of these animals may be humankind's most pervasive influence on nature."

In the aquatic realm, much of the study has focused on sharks and their potential to shape marine communities. As the exploitation of large sharks, in particular, intensified worldwide in recent decades, driven in part by an upsurge in demand for shark fins, the numbers of large predatory sharks sank like a stone. This led to populations of "mesopredators"—the predators below the apex level, which in this case may include rays, skates, and smaller sharks—quickly expanding.

For another study published in *Science* in 2007, researchers looked at shark populations along the US Eastern Seaboard and found that as the abundance of eleven great shark species there fell over the preceding thirty-five years, twelve of fourteen of their prey species increased. One of those prey species happened to be the cownose ray, a species of eagle ray, which eats a range of invertebrates that includes scallops. The enhanced predation of the rays on bay scallops, the study found, led scallop numbers to drop enough to terminate a century-long scallop fishery in the region.

Many other studies of trophic cascade involving sharks, as well as other marine apex predators, have been carried out, but far less attention has been given to similar impacts when it comes to freshwater species. It is, of course, yet another example of how understudied freshwater issues are. In this case, a large part of the problem is that we don't have a lot of good data on the abundance of freshwater apex species prior to their overfishing, so there's nothing to compare with.

What we can do is look at what has happened when apex predators have been introduced or reintroduced into a system. For example, when arapaimas first began to be used as an aquaculture species, they were placed into reservoirs where they hadn't been present before, and within a few years they had killed off much of the native fish diversity in the reservoirs. It's reasonable to deduce that if the introduction of the arapaima, or another apex predator, into a new environment will cause the massive reduction in other species, its removal would cause the opposite effect.

And so it often is with much of freshwater fish research. You have to take bits and pieces of information from other fields of study and apply them the best you can to your work. There is an urgent need to better understand the impact of the decline of freshwater apex predators on ecosystems. But we can look to what's happening with top predators on land and in the ocean. If we find ample evidence there of disastrous consequences, and we certainly do, there is little reason to believe the situation is better in our rivers and lakes.

~~~

While they aren't likely to eat you, there are many ways that an animal can hurt a human. Some will bite, while others claw. A python may on the rare occasion strangle you, and if you're not careful with a scorpion or a stingray, you might get stung. What these different modes of attack have in common is that the animal can harm you only if it can touch you. But there is one creature that can hurt you without making physical contact: the electric eel (*Electrophorus electricus*), an eight-foot-long freshwater predator widespread in the streams and ponds of the Amazon.

Despite its name and serpentine body, the electric eel is not a true eel but a member of the knifefish family. Unlike most megafish, which may look scarier than they really are (like the alligator gar), the electric eel is a fish to be genuinely careful with, because it has the capacity to deliver a serious electric jolt.

I had always wanted to get shocked by an electric eel. Not in a daredevil way, and not to get hurt, of course, but as a matter of curiosity. I had seen videos of people being shocked by them, and it looked uncomfortable but relatively harmless. The way I understood it, to be truly hurt, or to die, you had to be unlucky enough to fall into a pond with a hundred of these creatures or to take a direct and powerful shock to the chest, then fall into the water and drown. Still, it was with a mix of trepidation and excitement that I headed for the Amazon in 2014 with a National Geographic crew to film a *Monster Fish* show about the electric eel.

Once again we started our trip in Manaus, the Amazonian jungle city where we had begun our previous Amazon shoots. This time we did not visit the fish market. Instead, we headed to a nearby urban nature reserve, which we had been told was home to a group of especially large eels. It was going to be my first experience coming in contact with electric eels, and to handle them I had to wear insulating waders and thick rubber gloves.

Wading into a small pond, we corralled a large eel about six feet long onto a tarp. Even with the protective gear, I could feel the eel reacting with pulses of electricity as I handled it. I must have had holes in my gloves, because my hands got clammy and my fingers tingled as I felt a pulsed jolt each time I touched the eel.

Eels have been known to leap out of the water to deliver a so-called defensive electrocution, the most powerful shock dispatched directly to the body of a potential predator. While the average shock from an electric eel lasts only two-thousandths of a second, it maxes out at about 650 volts. (The voltage of most outlets in the United States is 120.) These numbers played in my head as I handled the eels. Feeling the tingling sensation through my gloves was enough to set off alarm bells in my subconscious. It was as if my primeval lizard brain was sending signals to my body saying, "Stop! This is not something you should be touching."

From Manaus, we traveled by boat down the Amazon until we reached the Tapajós River, a major tributary of the Amazon and one of the many rivers in the basin where electric eels are common. We were joined by Will Crampton, a biologist at the University of Central Florida and an expert on electric fish. Will had surveyed eels in the area and was eager to join our team for his own research.

Our boat was a three-story river cruiser, similar to the thousands of other cruisers that you see docked in the Amazon river cities, and for the next ten days it was our home. Painted bright white and blue, the boat was spacious, with open air decks on the second and third floors where tables and chairs were set up during the day and sleeping hammocks for the crew and film team to sleep in at night. I was actually given my own closet-sized room behind the captain's bridge, just big enough for a small cot. The room smelled of fresh paint and diesel fuel, but I accepted it gladly. During intense and exhausting field shoots, having your own space to escape to, however cramped, was always welcome.

Although electric eels are common, they're not necessarily easy to find. Many methods have been used throughout the years to catch electric eels. In 1800, the Prussian naturalist Alexander von Humboldt went on a four-month-long expedition to South America's Orinoco River to collect electric eels for experiments related to his interest in galvanism, the therapeutic use of electric current.

Locals had suggested Humboldt go "horse fishing," which entailed corralling wild horses and forcing them into the shallow water, where they would rile up the eels and compel them to attack by pressing their long bodies to the horses' bellies, releasing electric shocks. While many of the horses drowned, the effort was successful, according to Humboldt's account, as the eels quickly exhausted themselves and were easy to catch with small harpoons on ropes.

We were not going to try anything similar. Instead, in the flooded forests where we assumed the eels were hanging out, we deployed fyke nets. These contraptions consist of a long mesh-net wall that directs fish into a box-shaped trap. I had used them before, in Lake Tahoe, to catch trout.

As we waded waist-deep into dark, plant-choked stillwater swamps to set the traps, we unintentionally disturbed wasp nests. Now, this is not something you want to do in the middle of the Amazon, hours away from first aid and days away from a hospital. The first person to get attacked was our second cameraman, who was stung so many times he wasn't fully recovered for several days.

On another occasion, we must have accidentally hit a wasp nest with our small boat while fighting to get through a narrow channel with particularly heavy vegetation. It was past sunset and getting dark, when suddenly we heard a menacing buzz in the air. In the next instant, the producer behind me cried out in pain. Then, the other cameraman suddenly jumped overboard and dove under the surface to get away from the attacking wasps.

I could feel myself getting stung over and over, including on my head and on my neck. We scrambled and managed to get back to the main boat. By that time, one side of my face had swelled up, as had my neck, and I was concerned it would get worse overnight. We had brought at least one EpiPen, the device used for injecting the drug epinephrine into your body when you're having a serious allergic reaction, and I stayed up late into the night trying to decide whether to use it. Eventually the swelling subsided, and after taking a break from filming the next morning to recover and rest up, we were able to get back out the next afternoon.

Not that it got much easier. The fyke traps did catch electric eels, but they also caught a lot of other things. You had to be ready for anything. A catch could include swamp eels—true eels that are harmless, but not the focus of our work—and potentially venomous snakes, which in the shaded and murky areas where we were fishing were hard to distinguish from electric eels. The electric eel itself needed to be handled with gloves and great care—all while we were standing in waist-deep water in the dark, surrounded by mosquitoes and thorny bushes and mud, trying to film.

One day, Will took us out into a network of clear-flowing channels, bringing with him a device with which he said he could detect eels. It looked like a metal detector. Will would work his way along the channels with it, spending extra time around sunken logs and overhanging banks, while I followed him with a net, and Colm, the cameraman, followed behind both of us. I had no idea how the device worked, but it seemed to detect electric eels pretty much everywhere.

After several hours, I was ready to give up when, suddenly, out of the corner of my eye, I saw Colm slump into the water. I didn't understand what was happening, and neither did anyone else, but we rushed to grab his arms and pull him up before he went underwater, gear and all. We managed to get him to the bank of the channel and asked him what had happened. He was confused and had a hard time speaking. And that's when we realized that all the commotion in the water had disturbed an eel, and by bad luck it had bumped into Colm's leg. While Will and I had been wearing insulating waders, Colm had not, and he had received a jolt from the eel strong enough to leave him stunned and confused.

Our most successful technique of catching the eels involved tying strings with baited hooks to trees submerged in the flooded forest. We caught several eels this way, and it was safer than using the fyke traps, because we could handle the fish from the boat rather than standing in the water. Each time we caught an eel, we measured its length and weight, and after that Will measured the voltage generated from a shock. He did this by putting the fish on a nonconductive plastic tarp and placing electrodes on the fish's tail and snout. These electrodes connected to a voltage-measuring oscilloscope. Because electric eels are air-breathing fish, being out of the water for a few moments did not harm them.

At one point, we caught an eel and brought it to a nearby sandy beach where Will had set up his equipment. The eel was medium-sized, maybe three feet long, and, as usual, Will placed the electrodes on its nose and tail. As the fish tensed and shocked the gear, the energy pulse shot up to 860 volts. We thought this had to be an error. No electric eel shock had ever been measured at more than 650 volts before. So Will tried it again. But the result was the same.

We had caught a record-breaking fish. Not the kind I was usually after—a giant in size—but one that had generated the strongest electric shock ever recorded in any electric fish. It was a finding that ended up generating a lot of buzz (one could even say it shocked the research community), and the record was eventually entered into the *Guinness World Records*. But that was not all. Further analysis showed the record-shattering 860-volt eel was also its own species, separate from the one discovered 250 years earlier.

Will and other scientists studying electric fish had long suspected there was more than one electric eel species, and in the three years following our trip to the Amazon they collected 107 live electric eel specimens from across Greater Amazonia, testing their voltage and extracting muscle samples. An analysis of the morphological, genetic, and electric differences between the specimens revealed that they represented two distinct species in addition to *Electrophorus electricus*. It was the first time that electric voltage had been used to differentiate animal species, and the species with the record-breaking voltage came to be named *Electrophorus voltai* after the Italian electric battery inventor, Alessandro Volta.

The discovery added another piece to our ever-evolving understanding of these remarkable creatures. I was reminded of how little we know about all the megafish, or rather, how much there is to learn, and I couldn't help but wonder where that quest for knowledge would take us. In the case of the electric eel, it is a species that almost certainly will continue to spark interest.

Having learned how to swim when I was very young, I practically lived in the community swimming pool as a kid. But then, when I was six years old, *Jaws* premiered on television (to an audience of eighty million viewers), and everything changed. The next day, I wouldn't even dip my toe in the water. Not with the great white shark lurking at the deep end. Never mind that you could see the bottom of the pool. I knew the beast was there, waiting for me.

I was hardly the only one for whom the movie instilled such fear. Carl Gottlieb, who cowrote the screenplay for the Steven Spielberg-directed blockbuster, said that for years grown men and women that he met would tell him that after seeing *Jaws* they wouldn't even take a bath. The summer the movie came out, in 1975, beaches up and down the US coastlines sat empty. It's hard to think of a movie that has ever provoked such mass fear and hysterics.

What was it that scared people so much about *Jaws,* apart from its terrifying score? A lot of it surely had to do with the fear of what could not be seen, the unknown lurking below. I know that was a big thing for me. (After all, we never actually get to see what attacks the young woman in the opening scene.) But a lot of it, no doubt, had to do with the shark itself. This was not a fictional creature. It was an animal that existed in the real world. And people have always feared sharks. Unlike the mythological creatures that Richard Ellis wrote about in *Monsters of the Sea,* from the beginning the shark "retained the name and form with which it insinuated itself into human awareness." While the giant squid became the kraken and the octopus the popyp, the shark was always its own monster.

But as *Jaws* expanded our innate, if hugely misplaced, ideas of sharks as killing machines, the killing machine in many of us was also cranked up. Thousands of fishers, seduced by bounties and tournament prizes, set out to catch trophy sharks after seeing *Jaws.* Killings were proudly displayed with no remorse, since sharks were seen as nothing but man-killers anyway. George Burgess, a shark expert at the University of Florida, has estimated that the number of large sharks fell by 50 percent along the Eastern Seaboard of North America in the years following the release of *Jaws.* Later research by biologist Julia Baum at the University of Victoria showed that by the year 2000, populations of tiger sharks, great whites, and hammerheads had declined by 65 to 89 percent in the northwest Atlantic Ocean.

The damage done to sharks by sport fishing, however, pales in comparison to that caused by commercial fishing around the world. While sharks are usually not targeted by commercial fishers for their meat, they've long been killed as bycatch. Even more devastating has been the wanton slaughter of sharks for their fins, which are used to make shark fin soup, a dish whose popularity began to soar in Asia in the 1980s. Shark finning is a cruel practice, in which fishers cut off the shark's fins before throwing the fish alive back in the ocean, where, unable to swim properly and bleeding profusely, it will die a slow death.

According to a study in *Marine Policy,* by the year 2000 an estimated 100 million sharks were being killed indirectly or directly by humans each year, though the paper found that the number could have been as high as 273 million. And yet, amid those dire numbers, a movement to save shark populations also began to grow, much of it emerging in the United States but soon landing around the world.

As shark science proliferated, researchers made stunning new discoveries about the diversity and behavior of sharks and their importance to the ecosystem, and in time that knowledge seeped through to the public. Increasingly, people realized that sharks were not the vengeful killers portrayed in *Jaws*. In fact, only about ten people a year die in shark accidents worldwide, far fewer than are struck by lightning. Rather, sharks are some of the most incredible and important creatures on Earth, with an evolutionary history that goes back more than four hundred million years.

Strong protections for sharks were eventually implemented in many places around the world, including the United States, where congressional legislation in 2000 and the subsequent Shark Conservation Act of 2010 finally outlawed the shark-finning trade. Shark sanctuaries have become commonplace. Meanwhile, the success of Shark Week, the weeklong television programming that began on the Discovery Channel in 1988, has further stoked the public's fascination with sharks. While I appreciate that many shark experts have serious concerns about the more sensationalist aspects of Shark Week—having swimmer Michael Phelps race a computer-generated great white, for example, was probably not a great idea—it has helped pique curiosity about sharks and the threats they face.

Even if shark numbers, especially those for great whites, have rebounded in many areas around the world, there is still plenty of cause for concern and much work to be done. The Chinese government may have banned shark fin soup at its official banquets, but you can still find the dish served in thousands of restaurants throughout Asia. A *Nature* report in 2019 found that the global abundance of oceanic sharks (and rays, which were also included in the study) has declined by 71 percent since 1970, with many of the steepest declines in the North Atlantic, suggesting that many more shark species than previously believed are at risk of extinction.

But it is important to celebrate the progress made in shark conservation and learn from it. I have thought a lot about sharks—how and why the perception of them has changed. Like much of the public, I have gone through that change myself. Because of course I did get back in the swimming pool (in all honesty, it didn't take that long), and then I entered the ocean, and eventually, many years later, I found myself in the waters off Bimini in the Bahamas, surrounded by a group of hammerhead sharks while filming an episode of *Monster Fish*.

As we provisioned them (that's science speak for fed), the sharks swam in big circles around us, occasionally breaking away from their pattern and coming straight for the person holding the pole with the bait box. There were times when the sharks bumped into me, and once one of them accidentally bit into the cameraman's fin, only to quickly realize he wasn't food and move on. As someone who was scared of sharks as a kid, it felt like I had come full circle, thirty feet down in the crystal-clear water, surrounded by three or four large hammerheads, not afraid but amazed and filled with gratitude for the chance to be next to them.

The story about sharks has changed, from one of fearfulness to appreciation, and the same can happen for freshwater megafish. Once you understand that the sawfish will not use its sawlike bill to hurt you but to stun small fish, it isn't scary anymore. After you spend time with a playful goonch at the bottom of an Indian river, its beady eyes no longer look menacing. And a day of tagging the alligator gar, with its fearsome teeth and armor, will make you realize that the fish would much rather be left alone than cause any problems for bathers or fishermen.

I get that some of the megafishes are not what most of us would describe as pretty. (In fact, after my lectures, someone will inevitably feel the need to tell me that the fish I study are the ugliest animals he or she has ever seen.) But I know that sharks were once dismissed as ugly too, and now they're often called beautiful. And for every person making the ugly remark about my fish, there are ten others who tell me how incredible these animals are.

As for calling them monsters, I'm okay with that. Who doesn't love a good monster tale?

The Age of Loneliness

On November 14, 2003, as my interest in megafish was growing, the journal *Science* published a letter to the editor that caught my attention. It warned of the risks to ancient fish posed by the building of the world's largest hydroelectric dam, the Three Gorges Dam, on the Yangtze River in China.

The letter to the editor, written by Ping Xie of the Institute of Hydrobiology of Wuhan, identified three extremely vulnerable endemic fish species in the Yangtze that had already seen their populations sharply decline, and that were now listed as Grade I endangered species in the *China Red Data Book of Endangered Animals*. Two of them were sturgeons—the Chinese sturgeon and the Yangtze (or Dabry's) sturgeon—fish whose migratory routes had been almost completely blocked by another dam built downstream two decades earlier and now stood to lose their only remaining breeding ground because of the new Three Gorges project.

But it was the third species mentioned in the letter, the Chinese paddlefish (*Psephurus gladius*), that really aroused my curiosity, in part because of the undated photo that accompanied the text. It depicted a giant paddlefish propped up on rocks on a riverbank in Nanjing. With its elongated snout, sleek silver-and-gray skin, and red, streamlined fins, the creature had the look of a vintage sports car. A caption claimed paddlefish could grow to over seven meters (twenty-three feet) long, making it the longest freshwater fish in the world.

While "longest" didn't mean heaviest, the Chinese paddlefish was clearly a very large fish, and there was no reason to doubt the authenticity of a photo published in one of the world's most respected scientific journals. It seemed plausible that the Chinese paddlefish could in fact be the freshwater fish champion of the world. As the Megafishes Project got up and running, I considered it one of the top contenders.

The Chinese paddlefish is one of six known species in the paddlefish family, though four have gone extinct and are known only from fossil remains. The other extant species is another giant: the Mississippi, also called American, paddlefish (*Polyodon spathula*). Closely related to sturgeons—and resembling the sawfish in form—paddlefishes can trace their lineage back more than 150 million years.

In China, the paddlefish has played an important role both economically and culturally. Fishing of the species goes back centuries, if not millennia; there are stories of the fish being fed as a delicacy to Chinese emperors of the ancient past. With a snout that vaguely resembles an elephant trunk, it is sometimes called "elephant fish," and also the "panda of the river" because of its rarity and endangered status.

Just how endangered the Chinese paddlefish was, however, was difficult to ascertain. Almost nothing had been published scientifically about the species, at least not in English. From the bits and pieces I could gather, I learned that the fishery for the Chinese paddlefish had existed into the 1970s, with annual harvests reaching twenty-five tons. But the construction in the 1980s of the Gezhouba Dam across the Yangtze, a few miles downstream from the city of Yichang, had dealt a devastating blow to the giant fish, blocking off its main migration route, just as it had done to the two sturgeon species. After that, populations appear to have crashed.

In 1990, the Chinese paddlefish was red-listed as "critically endangered," while the Chinese government labeled it a "first-class state protected animal." The protections didn't seem to help much, however, because no young Chinese paddlefish had been seen in the wild since 1995, and there had been no sightings of wild Chinese paddlefish of any size since 2003. It became clear that the sparsity of information on the species wasn't just because I couldn't find data, but because the species itself had become so rare it was hard to study. Maybe it had even gone extinct, or was on the brink of doing so.

The question of whether a species has really become extinct is complicated in the context of conservation science. It's almost impossible to answer it with complete certainty, because—as the researchers who couldn't rule out a monster living in Loch Ness suggested—the absence of evidence does not mean evidence of absence. When do we say that a species has absolutely, positively, without a doubt lost every single individual of its kind in the world, and thus gone extinct?

In the case of the Chinese paddlefish, the situation was clearly urgent. The species was found only in the four-thousand-mile-long Yangtze River, which, coincidentally, has one of the highest rates of endemism of any major river. Almost half of its estimated four hundred species of fish are found nowhere else in the world. Other endemic fish had gone extinct in China in recent times—including two small ray-finned fish species that lost their only habitat, the Yilong Lake in Yunnan Province, when it temporarily dried up in 1981 because of water extraction. But as far as I could tell, no fish had been declared extinct from the Yangtze River, at least not recently.

There was, however, another large and endemic Yangtze animal that seemed to be heading in that direction: a freshwater dolphin species known as the baiji.

Also called the Chinese river dolphin, the baiji (the word means "white fin" in Chinese) held a sacred status in Chinese culture. Regarded as a symbol of peace and prosperity by fishers and boatmen, the population of the "Goddess of the Yangtze" had remained relatively stable throughout history. An estimate in the 1950s suggested that six thousand animals lived in the river, about the same number thought to have been in existence when the dolphin was mentioned in the ancient dictionary *Erya* in the third century BC.

As China embarked on its disastrous Great Leap Forward in the late 1950s, people began to kill their former dolphin protectors. Heavy use of the river—not just for fishing but for transportation and hydropower—caused the baiji population to decline dramatically, with a 1978 survey suggesting only four hundred animals remained.

At that point, conservation efforts were ramped up to save the baiji. As the crisis received more international attention, Chinese researchers set up plans for lake sanctuaries in which dolphins could be placed to establish breeding populations. But as the onslaught on the river environment escalated in the 1980s and 1990s, dolphin numbers continued to dwindle. And the few dolphins that researchers were able to capture did not survive. A survey at the turn of the millennium estimated thirteen animals remained in the wild. Then, in 2002, the one baiji held in captivity died.

A few years after that, some hope was injected into the conservation campaign to save the baiji when a Swiss businessman announced plans to survey almost the entire Yangtze River for it. But the plans were repeatedly delayed amid funding problems, and it was not until late 2006 that the expedition finally got going. By then, it was too late. The baiji had disappeared. For six weeks, the team of Chinese and international experts scoured the river without finding any signs of the dolphins, a failure captured in great and captivating detail by Samuel Turvey, a biologist, in his book *Witness to Extinction.*

The baiji had earned the sad designation of being the first dolphin species driven to extinction by humans, as well as the first large aquatic mammal to go extinct since hunting and overfishing had killed off the Caribbean monk seal in the 1950s. I wondered if the Chinese paddlefish would now become the first freshwater megafish to disappear in recent times. To find out, I went to China.

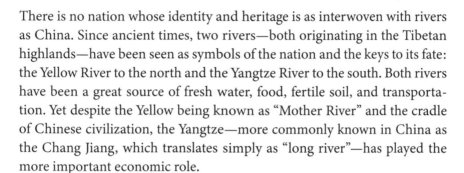

There is no nation whose identity and heritage is as interwoven with rivers as China. Since ancient times, two rivers—both originating in the Tibetan highlands—have been seen as symbols of the nation and the keys to its fate: the Yellow River to the north and the Yangtze River to the south. Both rivers have been a great source of fresh water, food, fertile soil, and transportation. Yet despite the Yellow being known as "Mother River" and the cradle of Chinese civilization, the Yangtze—more commonly known in China as the Chang Jiang, which translates simply as "long river"—has played the more important economic role.

The Yangtze River has been the backbone of transportation of goods for more than two thousand years, connecting China's heartland cities with the coastal city of Shanghai and by extension the outside world. The regions through which it runs, with a milder climate than that found in the north, are particularly suitable for agriculture, and several hundred years before Christ, during the Warring States period of China, a remarkable irrigation system was built along the Yangtze that provided stable farming for China and is still in use today.

But just as the Yangtze has provided immeasurable benefits, it has also wrought unspeakable destruction. The middle and lower basins of the Yangtze, which make up the most densely populated part of China, were originally a huge floodplain network that included the river and its tributaries and many interconnected shallow lakes. Over centuries, the Chinese constructed numerous flood-control projects designed mainly to increase farmland along the river, separating almost all of the lakes from the river and causing serious fragmentation of the system as a whole.

Nonetheless, catastrophic flooding still occurred regularly along the Yangtze well into the twentieth century. A series of floods that struck cities along the Yangtze over three months in 1931 caused the drowning of hundreds of thousands of people and many more deaths from flood-borne diseases. With estimated fatalities running as high as four million, the Yangtze-Huai River floods are often referred to as the deadliest natural disaster in human history.

The Chinese Communist Revolution, in 1949, would mark a new era for the Yangtze. When Mao Zedong concluded that "the south has plenty of water, the north much less; if possible, the north should borrow a little," the groundwork was laid for a massive water diversion project linking the Yangtze with regions in the north. Built over decades, it became known as the South–North Water Transfer Project.

By the 1970s eighty thousand dams had been erected around the country, primarily for irrigation. But the construction had been carried out mostly by unskilled peasants, with disastrous results. In 1973 alone, more than five hundred dams collapsed in China. Then, in 1975, the year before Mao's death, Typhoon Nina caused the Banqiao Dam in the Ru, a tributary of the Hong River, to give way, releasing a violent flood that devastated everything in its path. Nearly a quarter of a million people are estimated to have died as a result, handing China the unenviable distinction of presiding over not only the deadliest natural disaster in human history, but also the world's deadliest infrastructure failure ever (although the true death toll was never confirmed, as the catastrophe was kept a closely guarded secret for more than thirty years).

Mao's successors turned their focus to dams as a source of energy. The first hydropower plant on the main stem of the Yangtze, the giant Gezhouba Dam, came online in 1989, and later that same year—in the aftermath of the Tiananmen Square massacre—the final plans for what was to become the world's largest power station, the Three Gorges Dam, were pushed through the National People's Congress.

Throughout the 1990s and into the 2000s the hydropower boom continued, and by the time I arrived in China, up to fifty thousand dams had been built in the Yangtze River watershed alone. I was told there was only one river in the entire country—the Nu, which starts on the Tibetan Plateau and runs south, into Myanmar—that remained undammed, but there were plans to build a cascade of thirteen dams on it as well. Virtually none of the dam construction seemed to have taken into account the ecological consequences, leaving the country's rivers mere shadows of their former selves. And those were the ones that still remained. Of the fifty thousand rivers that existed in China in the 1980s, half had simply disappeared.

Just before the first official *Monster Fish* shoot in the Amazon, the World Wildlife Fund put me in touch with a Chinese scientist named Wei Qiwei, the leading expert on the Chinese paddlefish. I told him about my project, and he invited me to come to China to learn firsthand about the situation with the Chinese paddlefish. Stefan, who had come to Cambodia to cover my work for National Geographic's online news service, got a greenlight to do a series of multimedia stories on the Megafishes Project. So I invited Stefan to come to China with me and start the series there.

I also invited my brother Zack, who had lived in Taiwan for many years and spoke fluent Mandarin. His language skills became useful from the get-go, as we missed our flight from Shanghai to the city of Jingzhou, where Dr. Wei was based. Zack not only negotiated our way on board another flight, but he also made sure that the people we were meeting in Jingzhou knew of our late arrival. This turned out to be important, because when we arrived it became clear that our visit had been meticulously planned by our hosts, down to the minute, starting with a welcoming event attended by a large gathering of fish biologists and students. "It is an honor to welcome this important delegation," Wei told the audience to our surprise.

A bespectacled man in his mid-forties, Wei grew up on the banks of the De'an River, a small tributary of the Yangtze. The river, which Wei said had water "oh, so clear that you could see all the fish," ran through all his childhood memories. His grandfather had taught him how to catch mandarin fish ("the best-tasting fish in China") using trained cormorants with their necks tied so they couldn't swallow the fish before delivering it to the fisher. Then, Mao Zedong's Cultural Revolution in the late 1960s caused the family to be forcibly relocated to a small mountain village to do subsistence farming. Out of school, Wei spent three years herding cows and watching the forest around him disappear, with trees chopped down for firewood.

After another forced relocation and more years of hardship, the family could finally return home, but by then the De'an River had been dammed and turned into a trickle of stagnant slop. The destruction of nature he witnessed in his youth left an indelible mark on Wei, and when the time came for his university studies he chose biology, even though his excellent exam scores would have allowed him to pursue a degree in scientific subjects considered more prestigious in China, such as physics or chemistry.

After graduating, he got a job at the governmental Yangtze River Fisheries Research Institute. Two decades later, he led its research laboratory. "Many of my colleagues went to study in North America and Europe," Wei explained. "I had to stay and look after my fish."

"His" fish were the two very large species found in the Yangtze, the Chinese sturgeon and the Chinese paddlefish, and he felt a particular obligation to look after the paddlefish, since its status was so critical. To Wei, the Chinese paddlefish had always had an aura of mystery around it. Overfishing had cut down its numbers significantly even before the Gezhouba Dam was built, and it was not until 1985, when he started working at the institute, that he saw his first paddlefish.

As he studied them, his fascination grew. He learned that the fish's distinctive snout was lined with receptors that could detect subtle electric feeds, allowing its owner to act like an underwater sniffer dog hunting its prey, which included crustaceans, gobies, and minnows. But studying the fish proved highly problematic because whenever one was caught, it quickly died. "It is a very nervous fish and easily distressed," said Wei. "We had no knowledge of how to keep it alive."

In the wake of the 1986 US–China Nature Conservation Protocol, a research delegation from the US Fish and Wildlife Service came to China in the early 1990s to study the Chinese paddlefish and sturgeon. Wei had a lot of ideas about where the fish spawned, its movement and habitats, which he was able to explore with the help of the Americans. Gradually he learned more about the fish, figuring out how to keep it alive, and after the Americans left he got some funding from the Chinese government to continue his research.

But it was clear that the paddlefish were becoming increasingly rare. Wei had found juveniles up to the mid-1990s, but after that had not seen any young fish. Then, in December of 2002, an eleven-foot paddlefish was caught by fishers in Nanjing, not far from the coast. Badly injured, it was taken to a research station and put in a large tank, where, remarkably, it survived despite serious damage to its air bladder and fins. That is, it survived for twenty-nine days. Then, spooked by the commotion caused by researchers as they attempted to warm the tank's near-freezing water, the fish accidentally got stuck between a tank wall and a pipe and died soon after.

One month later, a similarly sized individual was accidentally captured near Yibin, far upstream from the Gezhouba Dam. The fishers who caught it in the early morning hours notified local law enforcement, who in turn contacted Wei, who immediately headed off on a six-hour journey by plane, car, and boat, giving instructions on the way to the fishers on how to handle the fish in a net cage until he could get there. On arriving, Wei immediately had to sew up a foot-long wound that the fish had sustained. He and his colleagues injected it with steroids and medicines. As the fish responded well to the treatment, they released it back into the river that afternoon after attaching two ultrasonic tags to its back, so that they could finally learn more about the paddlefish's movements.

For several hours and well into the evening the team was able to follow the fish's signal from a rented boat. But then, around 10:00 p.m., the boat struck a rock and broke its propeller. The team barely managed to float back to the riverbank. From there, they were still receiving the fish's signals, which reached about a hundred meters. In the moonlight, Wei could even see the paddlefish skimming the water's surface, something he had seen other individuals do for some unknown reason.

He knew he needed to get the boat fixed as soon as possible, but it was the eve of the Chinese Lunar New Year, and all the shops were going to be closed for the next three days. They had no choice but to wait. When they were finally able to get back on the water, the fish was gone. For days, they traveled up and down the river, almost one thousand kilometers, without ever hearing the signal again.

It was the last paddlefish that Wei had seen. But he knew he needed to be prepared if one popped up again. So he procured a rescue boat large enough to accommodate a tank on board and a big lift in the back.

We took this sixty-three-foot research vessel up to Gezhouba Dam. Although it was my first time on the Yangtze, I was well aware of the river's degradation and had steeled myself for what I would see. But as we departed under a hazy, gray sky, the river turned out to be more depressing than I had imagined. The Yangtze seemed like a waterway totally devoid of natural life, its banks consolidated with concrete. The murkiness of the water could not disguise its polluted state, with a wide assortment of trash on its banks. But what struck me the most was the heavy traffic on the river, with a never-ending procession of barges and cargo ships moving in both directions. The Yangtze felt less like a river and more like a highway. How could the Chinese paddlefish—or any fish, for that matter—survive here?

The Gezhouba Dam was a monolith that stretched almost a mile and a half across the width of the river. It cut the river almost exactly in half and, constructed with no passage for fish, made it impossible for the Chinese paddlefish to move from their forage ground in the lower part of the river to the spawning grounds in the upper watershed. And given their particularly sensitive spawning process, paddlefish could no longer reproduce in their native river.

Although it could go without food for a long time—maybe even several months, a survival mechanism that had served it well—the paddlefish sat on top of the food chain, feeding on large amounts of fish at a time. But because so little was known about the species, Wei didn't know what impact its potential demise would have on the fish assemblage that still existed in the Yangtze.

Our boat sat motionless on the surface of a river that had once roared and raged. The dam had proved devastating not only to the paddlefish but the Chinese sturgeon as well, built as it was on top of prime sturgeon spawning ground. The sturgeon, Wei explained, required a sloping gravel bottom with just the right velocity of water flow for the sperm and egg to mix. That topography had still existed below the dam even after its construction. But then an additional half-mile-long cement wall had been constructed— without any environmental assessment conducted—to broaden the navigation channel, wiping out the spawning ground. Wei said that because the construction had been done illegally, there was a chance it could be undone and for the sturgeon breeding ground to be reestablished, but that to me sounded more like wishful thinking.

The next day we drove by car to see the Three Gorges Dam, thirty miles upriver from the Gezhouba Dam. Its construction had just been completed, and the dam was operating at only 10 percent capacity. Once fully up and running, it would generate 22,500 megawatts of electricity, making it the most productive hydroelectric dam in the world.

Perhaps no engineering project in modern history had been as controversial as the Three Gorges Dam. Work on it, which started in 1993, had been delayed for nearly forty years as the government struggled to reach a decision on plans for the project amid huge economic and environmental concerns. In the end, it had cost at least $25 billion (but perhaps twice that) to build. At least 1.2 million people had been forced to leave their homes, with the 386 square miles of area composing the dam's reservoir submerging 140 towns, 13 cities, and 1,600 historical sites.

As for the environmental impact, a 2003 study estimated that forty-four fish species were at risk of extinction due to the dam. Ominously, it had also been built on top of old waste facilities and mining operations. Critics warned that the dam would increase river pollution in the stagnant water upstream, as well as earthquake and landslide risks in the region. No fish ladders or other mechanisms for fish to pass had been installed. "Too expensive," Wei said with a shrug, though work was continuing on constructing an entire side channel with an elaborate lock system that would lift or lower entire cargo ships from one side to the other.

When we visited a viewpoint, it was too hazy to see even halfway across the width of the dam. That made it difficult to absorb the enormity of a development that had required 510,000 tons of steel to build, or roughly that of sixty Eiffel Towers. At its maximum, the reservoir could hold forty-two billion tons of water, a mass so enormous that it would affect Earth's rotation.

I watched as throngs of Chinese tourists formally posed for pictures at the viewpoint. I could see the pride they took in their country having achieved such an engineering marvel; I remember similar photos my family had taken posing in front of the Hoover Dam when I was a child. There was no denying the potential economic benefits of the new dam. It would provide a rapidly expanding and energy-hungry China with 10 percent of its electricity needs, as well as allow oceangoing freighters to navigate 1,400 miles inland, from Shanghai to Chongqing, a city of twelve million people. It would conceivably help protect millions of people from the periodic flooding that still plagued the Yangtze basin. But at what cost?

Wei arranged for us to drive across the dam structure on a road that was accessible only to officials and staff. From inside the car I could not see the river below the four-hundred-foot drop; it felt like we were completely disconnected from the waterway. Later, we took a boat ride in the direction of Jingzhou and beyond that the megacity of Wuhan. Oil refineries and rusted factories lined the shores, and the occasional sand-dredging operation ripped up the riverbed. This is what had become of the river they called the Amazon of the East, with a basin that had once been lined with lush forests home to elephants, rhinos, tigers, tapirs, and gibbons. It seemed like China had very little to show for its rich biological past.

Yet Wei remained optimistic, certain that the Chinese paddlefish was still out there. The juveniles he had seen a decade earlier would still have twenty, thirty, maybe even more years to live. He believed the upper Yangtze might be a possible last refuge for the paddlefish. "That area has many deep pools and underwater caves where the fish can hide," he said. If he could just find a couple, he could breed them in captivity.

I didn't share his optimism about the paddlefish, but I understood and respected his determination to find it. As I started the Megafishes Project and set out in the world, I learned there were others, like Wei, who were dedicated to saving freshwater giants. And I understood that even if we failed, there was value in trying, value in learning about these animals, even if they were to disappear. Or maybe even more so because of that fact.

The start date of the Anthropocene (the geological epoch defined by human influence) is open to interpretation, but the fact that we have entered another mass extinction event is becoming indisputable. While earlier extinction events were caused by volcanic eruptions or asteroid collisions, people like the environmental writer Elizabeth Kolbert have termed this one an "anthropogenic extinction" because, this time around, humans are to blame.

According to a 2020 PNAS (Proceedings of the National Academy of Sciences) study, species extinction rates are today hundreds or even thousands of times faster than "normal." In the case of mammals, the fossil record indicates that the "background" rate of extinction—the one that prevailed before humans arrived—is so low that over the course of a millennium, only a single species should disappear. However, according to a study reported in *Science Advances* in 2015, at least 468 vertebrate species have gone extinct since 1900, including 69 mammals, 80 birds, 24 reptiles, 146 amphibians, and 158 fish species. But because these figures were arrived at using a highly conservative approach based solely on confirmed vertebrate extinctions, the real numbers are likely much higher.

According to the PNAS study, this surge in extinctions is happening for several reasons. For example, close ecological interactions of species on the brink of extinction tend to move other species toward annihilation when they disappear. In other words, extinction breeds extinctions. Even more importantly, human pressures on the biosphere keep growing and show few signs of letting up.

Starting in the Pleistocene era, which began 2.6 million years ago and lasted until about 11,700 years ago, mammoths and mastodons that had roamed the lands for millions of years disappeared in what could be considered the blink of an eye on the evolutionary timescale. Most experts agree that it was a combination of the animals being hunted by humans and other factors, such as climate change, that killed them off.

No such eradication happened in the seas, until people began to fish on a large scale. But in the last few centuries, we began to see a dramatic decline in the abundance of whales and large fish, too, which is attributable in large part to overharvesting, though it should be noted that few absolute extinctions appear to have occurred. As for freshwater fish, most species in trouble have seen their numbers plummet in the last half century.

In 2007, I collaborated with my colleagues Julian Olden and Jake Vander Zanden on a study of extinction risk, based on size, to the world's freshwater and marine fishes. For our study, which was published in the journal

Global Ecology and Biogeography, we collected maximum body length data for 22,800 freshwater and marine species. We found that large-bodied fish, in both the oceans and fresh water, are most likely to be at risk of extinction.

This was to be expected. But we also found that smaller freshwater fish were at greater risk of extinction than their marine counterparts. There is an obvious reason for this. While marine fish face one overwhelming threat, and that is overharvest (with large-bodied species disproportionately targeted), freshwater fish species of all sizes are subjected to a range of threats, not just overfishing but also habitat loss from dams, the introduction of invasive species, and other problems that are not as pervasive in the oceans.

On top of that, freshwater fish are confined to the rivers and lakes in which they live in a way that marine species are not. In many cases, freshwater fish are found in one place on Earth. When that river or lake is degraded or destroyed, the fish has nowhere else to go.

What about megafauna specifically? Well, population declines have been well documented in terrestrial and marine megafauna. But few studies had looked at global trends of large freshwater species. This started to change around 2010, when documentation of the disappearance of species like the Chinese paddlefish and the emerging data on the decline of giant catfish and barb finally stimulated more and larger studies.

This culminated in research that I was involved with and that was published in the journal *Global Change Biology* in 2019, where we presented grim findings: since 1970, the global populations of large freshwater animals—those weighing at least thirty kilos, or sixty-six pounds, and not just fish—have declined by almost 90 percent, more than twice as much as the loss of vertebrate populations on land or in the oceans. The numbers were even starker for large fish, whose numbers had plummeted by a staggering 94 percent.

How do you tell such a story? How do you make such numbers meaningful? How did I want people to respond to the findings? In an interview with the *New York Times,* I said: "We want to go beyond just studying conservation status and look at ways to try to improve the situation for these animals." I tried to sound hopeful. When the article came out, I read the reader comments on it. While several people were encouraging and expressed an interest in "doing something," others were decidedly more resigned. "There is no hope," wrote one particularly downbeat reader. "All the animals will be in zoos or dead. Face it."

What does it mean when a species or a population vanishes? It means our entire ecosystem erodes to a degree. Each species plays a role, and some, like the megafish, may play a more ecologically important role than others. What would happen if all the river giants died? Or ended up in a zoo? No one knows yet, but we are finding out, faster and faster, as we enter what the famed biologist E. O. Wilson dubbed, rather than the Anthropocene, "the age of loneliness."

One of the two sturgeons referenced in the *Science* article about the Three Gorges, the Chinese sturgeon (*Acipenser sinensis*), is an anadromous fish that can grow up to sixteen feet long and weigh thirteen hundred pounds. Although it is not an obligate freshwater fish, the story of its captive breeding in China and the environmental challenges the species faces are highly relevant to the discussion of megafish persistence in the Anthropocene.

With its sharklike form, large pectoral fins, and rows of pronounced ridges running the length of its spine and flanks, the Chinese sturgeon is a cool-looking fish. Known in China as the "mother species," it has a history and status as storied as its relative the Chinese paddlefish, playing a big role as a food fish in dynasties dating back more than a millennium.

Historically, the Chinese sturgeon swam across East Asia, including China, Japan, and the Korean peninsula, but it has long since been extirpated from all regions but the Yangtze River, where it was barely hanging on when I got there. It's a story of decline similar to that of the Chinese paddlefish. While the Chinese sturgeon had been overfished in the Yangtze, too—leading Chinese authorities to implement a ban on its fishing in the early 1980s—what really hurt the species was the construction of dams on the Yangtze, and in particular the Gezhouba Dam.

No longer could the sturgeon complete its two-thousand-mile journey—the longest migration of any sturgeon in the world—from the sea to its expansive breeding grounds in the Jinsha River, as the upper Yangtze is called. Instead it was left with an area of about seven kilometers below Gezhouba to breed, and even that area soon proved unsuitable. Historically, the sturgeon would have a long, progressively colder swim in which their bodies slowly shift into reproductive mode. Now, they had to jump into mating without all the physiological foreplay they had evolved in a place where water temperature, flow velocity, and other environmental parameters had all been altered. Not surprisingly, it wasn't working, especially since the Chinese sturgeon is particularly sensitive to such environmental cues.

The Chinese sturgeons do, however, have an advantage over the paddle-fish that will help them from going extinct: they can be bred in captivity. China has a long history of artificial sturgeon reproduction, going back to the 1950s when the first sturgeon species, the Amur sturgeon, was bred successfully there. This was followed in the 1970s by the breeding in artificial conditions of both Yangtze and Chinese sturgeons to be restocked into the Yangtze River.

Dr. Wei told us that his institute operated a Chinese sturgeon breeding facility, and a few days into our visit he invited us to see it. "Love the Chinese Sturgeon, Our National Treasure," read a slogan on the wall as we entered the compound, which was tucked into a sleepy farming community outside Jingzhou. Inside, a space the size of two football fields contained rows and rows of tanks holding sturgeons in varying stages of development, from larvae to one-year-old fish, with sexually mature sturgeons taken from the wild kept there to provide eggs for breeding.

It was an impressive operation and a worthy endeavor. But I was not surprised to hear that, so far, there had been no evidence of reintroduced Chinese sturgeon spawning in the wild.

There were also concerns about hatchery fish diluting the genetic diversity of wild fish. I had seen this play out in Thailand, where the Department of Fisheries had developed methods in the 1980s to breed Mekong giant catfish, an accomplishment that likely saved the giant catfish from extinction. While some of the captive stock had been released into artificial reservoirs, where they grew big but did not reproduce, other stocked juveniles had been released in the Mekong at sites that were almost certainly spawning habitat for their wild cousins. The fear was that these stocked fish with their poor genetic diversity—an analysis in 2001 showed that 95 percent of ten thousand giant catfish fingerlings released shared two parents—would overwhelm the small natural population and make the already rare catfish weaker and even less able to adapt to the changing conditions in the Mekong.

It was a similar scenario with the Chinese sturgeon in the Yangtze River, and Wei was well aware of such long-term problems. But like other hatchery proponents, he felt he had no choice but to forge ahead; the stakes were too high, the sturgeon could not be lost. Besides, he and the other researchers were dealing with a more urgent problem: the adult sturgeons at the facility were getting sick. It seemed like they were developing a bacteria problem in their stomachs. Wei thought it had something to do with the fish not having enough space to move.

He took me outside to see the adult sturgeons, which were kept in a series of cylindrical cement structures according to their size. The largest structure, about twenty-five feet across, housed two adult sturgeons. They were six or seven feet long—large, but only about half the size that a Chinese sturgeon is capable of attaining. As we watched them, they stayed side by side in the water, not making a move.

As we left the hatchery, I was filled with mixed emotions. I admired Wei's can-do spirit. He was convinced the Chinese sturgeon could survive. "In the future, people will be able to say that we were the ones who saved the Chinese sturgeon," he said. I wanted to share his optimism about the future of the species. But looking at the sorry state of the Yangtze River, it was difficult to imagine that any fish put into that water could survive. It was the same story as with other efforts to breed fish to repopulate rivers. Unless the ugly river conditions that caused the fish to disappear in the first place were improved, there was little hope of success.

As for the Chinese paddlefish, the outlook was even bleaker. There were no paddlefish kept in captivity, and the only specimen I had seen was a stuffed one in a museum. Wei, the eternal optimist, was still convinced it could be found, however, and so it came to be that at the goodbye banquet hosted for us back in Jingzhou, we signed a memorandum of understanding to embark on a future joint expedition, continuing the search for the elusive paddlefish.

From the beginning, my aim was to gather data on giant freshwater fish around the world; assess their conservation status and threats; and identify solutions to aid their survival. It was a mission that was, in many ways, in line with what the IUCN's Red List of Threatened Species was trying to achieve with all the animals and plants on Earth.

The Red List, also known as the Red Data Book, was founded in 1964 as the first comprehensive inventory of all biological species and their conservation status, and it has long been recognized as the best measure of how the world's wildlife is faring. The database has both grown and changed a lot over the years. Today, species are classified in seven categories, from "Least Concern" to "Extinct" (with two additional categories of "Data Deficient" and "Not Evaluated"). The criteria for evaluating species include their rate of decline, population size, area of geographic distribution, and degree of population and distribution fragmentation.

By the turn of the millennium, fewer than twenty thousand species had been assessed, with the numbers heavily skewed in favor of mammals and birds. (By 1988, virtually all bird species had been evaluated.) In the early 2000s assessments picked up speed, and by 2004 the total number from four years earlier had doubled. With about 40 percent of the assessed species described as threatened with extinction, the extent of the growing global biodiversity crisis was plain for all to see. "The library of life is burning," the Norwegian prime minister Gro Harlem Bruntland told an environmental conference, "and we don't even know the titles of the books."

In 2003, I was asked to evaluate the Mekong giant catfish for the IUCN project. Up to that point, and for some time afterward, both freshwater and marine fish species remained poorly represented on the list, except for sharks, for which a third of the species had been assessed by 2004, with 18 percent classified as threatened. It was already becoming clear, however, that freshwater species were among the most vulnerable animals in the world.

The Mekong giant catfish was actually one of the few large fish species that had been assessed, first as "vulnerable" in 1986, and in 1996 moved to "endangered." But the conditions affecting the species were rapidly deteriorating, and I concluded that, according to the assessment criteria, the Mekong giant catfish should be considered "critically endangered," the classification before "extinct." This fish was clearly at risk of disappearing in the wild. Having that documented and validated by the Red List, where the giant catfish was now listed alongside big, charismatic land animals that everyone had heard of, seemed symbolically significant, as well as urgently needed to spur conservation action.

In the years to follow, I continued conducting assessments for other freshwater fish species, including Mekong fish like the striped river catfish ("endangered") and the giant carp ("critically endangered"). With more freshwater fish, including the giant species, being assessed around the world each year, the list proved extremely useful in my research and also communicated the urgency of the biodiversity crisis to the public.

Yet the Red List also has its limitations. Good data are hard to find, assessments can quickly become outdated as environmental pressures change, and the list does not differentiate between different populations within a species. So while the white sturgeon in the Kootenai River, for example, would probably be considered "critically endangered," the species as a whole is judged to be of "least concern" on the Red List.

The biggest gripe that people seem to have about the list is that by focusing on extinction risk as the only metric of conservation status, it is missing the actual point of conservation. Avoiding extinction of a species is not enough, we want it recovered and thriving. The list, some people argue, makes us think of success in terms of what we want to prevent rather than what we want to achieve.

I can see the value in that argument, and so, apparently, can the IUCN, whose members, in 2012, resolved to address the problem by creating a separate list that celebrates species recovery and conservation success. I'm all for the new, aspirational list, but the most important thing when it comes to biological conservation is action, and real action is tough to pull off.

But real action has never been more critical. To date, more than 134,000 species have been assessed for the IUCN Red List, with 28 percent of them threatened with extinction. Over the last ten years, experts on the IUCN Species Survival Commission, meanwhile, have mapped and assessed the threats for more than twenty-five thousand freshwater species. Almost a third of these are threatened with extinction, and over two hundred are already believed to be extinct. As it turns out, one of them, which is now listed as "extinct" on the IUCN Red List, is the Chinese paddlefish.

〜〜〜

After I left China, Wei continued his search for the paddlefish. In the years that followed he conducted eight surveys, each spanning several weeks. He and his crew set thousands of fishing lines and laid nets up and down the Yangtze River in hopes of catching the elusive *P. gladius,* and several times he made acoustic contact with what he was convinced was a paddlefish, but he never saw one.

The news about fish, rivers, and the environment that was coming out of China in those years all seemed earth-shatteringly depressing. In 2007, soon after our departure, the Yellow River dried up and failed to reach the sea, its lower stretches reduced to a broad highway of cracked mud. More than 30 percent of fish species in the Yellow River had gone extinct, Chinese officials reported, while behind the Three Gorges Dam in the Yangtze, the weight of the water building up was causing regular seismic disturbances that had forced the relocation of another fifty thousand people.

In 2010, the Chinese sturgeon was listed as critically endangered. A couple of years after that another dam, Xiangjiaba, came online upstream from the Gezhouba. It was followed the next year by yet another dam on the Yangtze, the Xiluodu. Possible effects of these dams on the sturgeon hadn't been given much consideration, since the sturgeon's path to upstream areas had already been blocked off by the Gezhouba. But a report showed that the newer dams would have a compounding impact on the water temperature in the river, which could spell final disaster for the sturgeon.

In 2015, Stefan went back to China and met up with Dr. Wei. By that time, Wei's institute had undergone great change, with half of its staff moving away, most of them abroad. His office had relocated to the megacity of Wuhan. Wei was despondent. He said he felt tired. People were writing a lot of papers, he complained, but nothing ever came out of it. "Everyone is just floating on the surface," he said, which was a curious thing to say.

Yet he had not given up on finding the paddlefish. In fact, he was convinced there were up to a dozen individuals left in a deep pool refuge in the upper reaches of the Yangtze. But at that point, his once-shiny research vessel had not left the dock for some time. As it gathered dust, it seemed to have become permanently attached to the other boats and barges tied together at the riverbank.

Soon after that, an official survey found 1,864 giant pandas in the wild, while a separate report estimated that only about a hundred Chinese sturgeon remained in the Yangtze River. It was said that six million sturgeon, a staggering number, had been released into the river since the 1980s, but apparently none of them had survived. Then, in 2018, another news report came out quoting Dr. Wei: "It is probable that the Chinese paddlefish is now extinct, a situation which experts are reviewing."

In that year and the next, researchers embarked on a basin-wide biological survey to locate any individual paddlefish remaining in the Yangtze. Using various types of nets, sonar, electrofishing gear, and other techniques, they came up empty-handed and later declared that the species had indeed gone extinct, most likely between 2005 and 2010, which was during the time of our visit with Dr. Wei. The paddlefish had probably become functionally extinct—that is, unable to reproduce—many years before then, the researchers wrote in their report, which generated a lot of news coverage around the world.

I thought back to the baiji and what Samuel Turvey had written in *Witness to Extinction*. His conclusions about dolphin extinction could describe what had happened with the Chinese paddlefish and its demise. "Unlike all of the historical extinctions of other large, charismatic mammal and bird species…the disappearance of the baiji had not come about because of active persecution," he wrote. "Instead, it was merely the result of incidental mortality caused by huge-scale fishing efforts and development projects that were not actively trying to target the dolphin at all." What had really killed the baiji, he argued, was simply the progressive ecological deterioration of the Yangtze, and that is what had now also claimed what might very well have been the world's largest river fish.

Under Father Sky

The first time I traveled to Mongolia was in the fall of 2003. It was the year before I finished my PhD, and I was part of a group of scientists who were going to assess the impact of gold mining on Mongolian rivers. The project was led by Sudeep Chandra, a whip-smart and tireless limnologist who had been my housemate at UC Davis and was now a postdoc at the University of Wisconsin. Sudeep had visited Mongolia the year before and become fascinated with the country, its unique culture and natural beauty. Though "beautiful" is not the word I'd use from my initial impression of Ulaanbaatar, the Mongolian capital, as we flew in from Beijing.

More than a decade had passed since Mongolia had emancipated itself from the Soviet repression that had shackled the country for much of the twentieth century. But the city, sprawled across a river valley, still resembled a Soviet outpost. Huge smokestacks pierced the skyline above endless rows of concrete apartment buildings. Giant pipes crisscrossed the grounds of shantytowns filled with hundreds of identical yurts, or gers as they're called in Mongolia.

Ulaanbaatar had the feel of a Wild East, with foreign mining companies scrambling to get into Mongolia to exploit its earthly riches, including coal, copper, gold, silver, iron ore, zinc, and uranium, valued at up to $3 trillion. A lot of new money attracted a lot of Mongolians to the city, which now housed more than half of the country's three million people. With the Soviets departed but Russia still a power to the north and China encroaching from the south, Mongolians were eager to assert their independence. And while the nation was still finding its democratic feet, the enthusiasm and pride the Mongolians we encountered showed in their newfound political freedoms and future made a strong impression on me.

Apart from Sudeep and me, several other people in our group had UC Davis connections, including Brant Allen, a fish biologist and good friend of mine; Andy Stubblefield, a jovial grad student; Jim Thorne, who was an expert on maps; and Mimi Kessler, whose doctoral thesis focused on the great bustards of Mongolia. Also with us was a Wisconsin student named David Gilroy, who had spent a lot of time in the country and spoke the language fluently.

David was friends with a Mongolian fish biologist who joined our team, Erdenebat Manchin, whom we ended up calling Fish Erdenebat, since Erdenebat was a common name in Mongolia. Trained in the Soviet era, Fish Erdenebat was a knowledgeable and lovable character who was always very happy, especially when we were drinking and talking about fish, which happened almost every night.

After spending a few days in Ulaanbaatar, or UB as people called it, it was time to get moving. With the group split into two Russian-made vehicles— one a jeep and the other a van, both dubiously described as "executive" models—we set off to the east toward the Khentii Mountains. These mountains, I was told, were the birthplace of Genghis Khan—or Chinggis Khan, as he's known in Mongolia—the thirteenth-century Mongolian ruler who founded the largest contiguous land empire in history. I had seen his name and image emblazoned on seemingly every product and shop in the capital. While Genghis Khan has a reputation in the West as a fearsome conqueror, he is viewed by all Mongolians as a unifying leader and national hero.

Soon we were swallowed up by the vastness of the Mongolian land. With the winter coming, the thick stands of birch draping the mountainsides were turning yellow. Apart from the herders on horseback moving giant flocks of sheep across the steppe grasslands, I didn't see a lot of people. But then Mongolia is the least densely populated country in the world. Fish Erdenebat talked about the intimate connection his countrymen maintain to their land. The Mongolians had a saying, he told us: "We are the children of father sky and mother earth."

But that sacred view of nature was increasingly being challenged, and, as we would find out, it did not extend to those engaged in the mining boom. Our project had grown out of work that Sudeep and others had been conducting in Siberia's Lake Baikal, the world's largest freshwater lake by volume. Deteriorating environmental conditions there, specifically an increase in phosphorus levels, were suspected to be linked to mining operations in Mongolia, where most of the Selenge River basin, the largest single inflow to Lake Baikal, can be found. Our goal was to investigate the impact of gold mining, specifically, on the water quality of the many rivers throughout the Selenge basin, which as a whole is larger than the state of California.

The first part of the trip was very buggy, with swarms of gnats buzzing around in the various camp sites where we stayed. When it got really bad, we retreated to our vehicles, to read or talk while clouds of small flies hovered outside. At night we camped or slept in run-down hotels, while visiting different sections within the Yeroo watershed in the Khentii Mountains. Some of the tributaries were among the most pristine I'd ever seen, with clear water flowing through magical forest settings. But we also encountered places where entire riverbeds had been ravaged by placer mining, a practice in which gold is sifted out of the gravel on the river bottom by a tumbler inside a dredge, with the resulting gravel deposited behind it.

At one point, we came across a group of miners dredging a whole section of a river. The legality of the operation was questionable, especially since much of the region was supposed to be part of a protected reserve. Yet the miners didn't care that we saw what they were doing. I assumed they had nothing to fear from law enforcement. Neither, presumably, did the scores of "ninja miners" we encountered. These were Mongolians who were digging for gold using only small tools. Their nickname came from the green panning bowls that many of them carried on their backs, which resembled the shells sported by the Teenage Mutant Ninja Turtle cartoon characters. There were tens of thousands of ninja miners in Mongolia. Many of them had been traditional herders, who had turned to mining after losing their livestock during particularly harsh winters.

While the water quality assessment was the main focus of our study, there was a very interesting fish component to it, which is why I was invited along. The fish in question was not just any fish but *Hucho taimen,* the world's largest salmonid, the group of fish that includes salmon, trout, chars, freshwater whitefishes, and graylings. A strictly freshwater fish, the Siberian taimen, as it's also known, is one of five taimen species and one that by the turn of the millennium had been largely extirpated from much of its range, in China and Russia. Northern Mongolia had become one of its last strongholds. But it was clear the mining activities in Mongolia were negatively impacting spawning grounds for the taimen and could spell catastrophe for the fish.

I was curious about the taimen—not just because it was said to grow up to six feet long, but also because of its mythical aura. It was a fish that was revered in Mongolian Buddhism as the child of an ancient river spirit. The stories I heard described taimen hunting in packs, earning it the nickname "river wolf." Its appearance dramatically changed as it grew bigger and developed large, muscular shoulders, its head becoming compressed while the gape of its mouth increased. The fish, I was told, would eat anything it could fit into its mouth: muskrats, ducks, even bats. There were even stories of taimen eating their own. I wondered if this cannibalistic behavior, if true, was part of the reason they grew so big.

Chinese miners we encountered were catching a lot of fish to eat, mostly smaller lenok and grayling but also taimen. In contrast, fish had not traditionally been a major part of most Mongolians' diet or livelihoods, which was one reason why taimen populations in the country had remained relatively healthy. An exception was Fish Erdenebat, who would go out to fish with a Russian guy named Boris who had joined our group.

Erdenebat believed the best time to catch the taimen was when the moon was full. Erdenebat and Boris would take their homemade mouse lures and fishing line wrapped around a thick branch to handline in the most promising pools. Using an impressive technique, Erdenebat would swing the line with one hand, cast out, and slowly bring the lure back to him, his movements mimicking the twitches and splashes of a small rodent struggling in the water.

The two of them managed to catch a few fish of good size, at least compared to the size of trout that I was used to catching, which were maybe twenty-four inches. They wrapped the fish in a package of birch together with wild onion and cooked it over the open fire, resulting in a tasty late-night meal.

As impressive as those fish were, it was clear the taimen could grow a lot bigger. The bars and restaurants that we visited had enormous taimen heads displayed on the wall, often with beer cans placed vertically in their mouths. Boris also told us about a history museum in the provincial capital of Öndörkhaan located farther south that he said held a preserved taimen 1.67 meters long, or almost five and a half feet. A taimen of this size would approach the mythical two-meter mark, a size reported from old stories but never substantiated. Naturally, I was curious to check it out.

The town, located on the northern bank of the Kherlen River, turned out to be a pleasant place of tree-lined streets and scattered Genghis Khan monuments. It was home to several museums, including the one that was supposed to house the taimen, the Khentii Aimag Museum. When we arrived, however, the museum, an old and dusty building, was closed. Luckily, a small-statured caretaker of advanced age, who sported a superb Fu Manchu mustache, materialized to allow us inside.

The collection had not been spruced up in a long time, but the artifacts on display were nonetheless pretty nifty: there was a mastodon tusk, a protoceratops skull, lots of stuffed local wildlife, and an assortment of Chinggis-era armor. We did not, however, see any signs of a giant taimen, so we asked the caretaker about it. He disappeared into the bowels of the museum, emerging a moment later with a stuffed fish. The specimen, long and skinny, had not been preserved well and it was rotting and falling apart. But I could tell it was a salmonid. It was also bigger than the caretaker, almost as tall as me, and longer by far than any trout I had ever seen. This fish story, at least, was accurate and lent credence to old reports of even bigger fish.

From Khentii, we traveled hundreds of miles west to the Selenge River itself. If the environmental destruction we had witnessed in the Yeroo watershed was hard to bear, it was nothing compared to what awaited us at the Zaamar mining field in the province of Töv. There, giant dredges one hundred feet long and five stories tall, the shape and size of old-fashioned Mississippi steamboats, were demolishing entire river sections. Where the water had once flowed full and clear, only a series of giant mud pits remained, separated by huge heaps of desiccated gravel that must have once been spawning beds for taimen. Few of us had seen anything so egregiously destructive. It was a real gut check, and a realization that Mongolia's vast landscapes and pristine rivers might not remain so forever, as exploitation of the country's rich natural resources accelerated.

During the several weeks in the field, the weather had turned progressively colder, and when we returned to UB the temperature had dropped well below freezing. (Of all the world's capitals, UB has the largest temperature swing between summer and winter.) We had a few days left before flying home and were looking for a last excursion. Fish Erdenebat had told us that we could see taimen in the Upper Tuul River, a few hours' drive from the capital. Brant and I had brought wetsuits and camera gear with us for just such an opportunity, and so we decided to go and have a look.

When we got to the river, which was about fifty feet wide, we found that ice had already formed around its edges and in backwaters and sloughs. The water itself was dark and slid past under a gray sky filled with light snow. It did not look particularly inviting for a swim, but we both had experience diving in cold water in Lake Tahoe and weren't going to chicken out. After putting on our wetsuits and taking a few swigs from the bottle of vodka we had brought with us, we took the plunge. Ice-cold water immediately made its way inside my suit and down my back, spreading over my skin. It was not the most comfortable feeling, but it was followed by a rush of adrenaline, and I acclimated to the cold as we made our way to the middle of the river where we began drifting along the surface while searching for life below.

At first, we didn't see anything but the empty gravel bottom. After drifting maybe three or four hundred yards over shallow riffles, we began thinking the trip had been in vain. But then, at a bend in the river, we reached deeper water. The gravel bottom dropped away below us, and we were suspended over the dark water of a pool at least eight feet deep. Suddenly there were taimen everywhere, at least a dozen, maybe more, swimming right below us. They seemed to be of different sizes, two feet, three feet, even four feet long. There were dark shapes farther below that hinted at some true monsters.

Our thick wetsuits kept us buoyant, and we were unable to dive, so we could not get down to the deepest waters where the biggest fish appeared to stay. We brought Nikonos underwater cameras, which had arms on their sides with strobes to light the dark scene, and began to snap away. I was so excited. No one to my knowledge had ever taken good photos of the world's largest salmonid in its natural habitat.

I thought for sure that our pictures would be worthy of publication by *National Geographic*. But when I got back to the United States and developed slides of the photos we had shot (this was before the digital age), I was sorely disappointed. Because we had been unable to keep steady in the flowing river, and unable to dive to the deepest waters to get close to the big fish, all you could see was dark water and a lone fish somewhere far below poorly illuminated by the lights. We still sent our photos to the editors at *National Geographic,* but they were not impressed.

That first trip to Mongolia made a big impression on me. I had traveled to many countries around the world, but Mongolia was one of a kind. Breathtakingly beautiful with clean, clear air and wide-open spaces, the country had a wildness to it that was oddly reminiscent to me of other remote landscapes that looked nothing like it, like the Okavango Delta of Botswana and the hidden core of the Amazon rainforest.

It was easy to appreciate the connection to nature that Fish Erdenebat had talked about. My affinity for the place, and desire to go back there as soon as I left, also had to do with the pivotal moment in which the country found itself. The nation was undergoing a dizzying political transformation, and there was a lot at stake. With the mining assault on the country, the need for environmental action and conservation was urgent, and that clearly included protection of the world's largest trout.

At the end of our trip, in UB, we had met with Jeff Liebert, who worked with the International Finance Corporation, a sister organization of the World Bank. An economist by training, Jeff had been a Peace Corps volunteer in the Mongolian town of Mörön some years earlier and fallen in love with the country. He became friends with Dan Vermillion, one of three brothers who owned a Montana-based fly-fishing company called Sweetwater Travel, which in the late 1990s began organizing fly-fishing trips to Mongolia. Dan had taught Jeff how to fly fish for taimen and got Jeff hooked.

Now Jeff had an idea, which he had concocted with Dan's help. He was looking to create a community-based conservation program for taimen in the remote Eg-Uur watershed in northern Mongolia, in an area known as the Wild District. The idea was to establish a recreational catch-and-release fishery for taimen through the licensing of fishing concessions, which would raise tourist revenue for local communities and conservation initiatives to protect the taimen. For it to work, Jeff said, a scientific knowledge base about the taimen had to be built from scratch, since no research about the ecology of the fish had ever been done. What he was looking for was a group of young, eager, and intelligent scientists who could do this with funding from the World Bank.

It sounded like a dream project. Back in the United States, Sudeep and David immediately began work on a proposal together with Jake Vander Zanden, another postdoc at UC Davis who would later become a faculty member at the University of Wisconsin. That proposal was accepted, and it was agreed that we would receive $250,000 for a five-year project, which at the time sounded like a lot of money. I later realized that a quarter of a million dollars would barely support our team over several years, and that a willingness to carry out the work on a frugal budget had probably also been an important consideration for Jeff.

The proposal outlined a research program that would provide a scientific basis for the sustainable fishing of taimen. There were basic questions to be answered: How many fish were there in the Eg and Uur Rivers? Were the fish moving in and out of the area? Would there be incidental mortality from fishing for them? How long did they live? How fast did they grow? This was all information needed to manage a healthy fishery.

The proposal also had a faith-based element to it. It was important to connect the project to Mongolia's culture, especially since Mongolian

Buddhist tradition eschewed fishing. Sudeep and David came up with an interesting visual based on the traditional Buddhist prayer wheel to explain different parts of the project. There were practical applications too, including the proposed restoration of a local monastery that had been destroyed during the century of Soviet control, when Buddhism had been completely banned in the country.

A nonprofit organization called the Taimen Conservation Fund, headed by another Erdenebat, was set up in UB. Meanwhile, Sudeep and David traveled to Mongolia in the late spring of 2004 to begin the work of establishing a science camp in the Eg-Uur area. They were joined by an older Mongolian man named Bazarsad Chimed-Ochir. A former field operations manager for the World Wildlife Fund, Chimed had experience working in remote regions. But this was going to be a more challenging enterprise given the lack of precedent for using concessions to protect fish.

The site selected for the camp, along the Uur River, had no infrastructure to plug into. It was situated on a plot of well-grazed grass that belonged to local herders, who gave permission for the campsite on the condition that we would not bring or graze any livestock of our own. Getting there involved an eighteen-hour-plus van ride from UB, of which the last several hours were driven across open steppe or on narrow dirt tracks. It was decided the camp needed solar power, so Sudeep was tasked with procuring the panels to set one up, which turned into a logistical nightmare. There was also a lot of political wrangling with local leaders, made all the more difficult by the fact that no phone coverage existed. A tower was built for a radio communication system, but it rarely worked.

Back in the United States, I didn't know much of what was going on since I was not in direct contact with Sudeep and David. But when they finally returned home later that summer, they were both full of enthusiasm. It had been a chaotic few months, but now the bones of the camp had been put in place and it was time to get the rest of the team over there to start the research in earnest. And so, in the fall of 2004, almost exactly a year after my first trip to Mongolia, I found myself back in the same Russian van, this time making the journey north toward a remote corner of Mongolia. In the slow and bouncy van, it felt like traveling to the end of the earth.

The setting for the camp was next to a pristine river winding its way through the wide valley. The actual camp consisted of a single ger, perhaps fifteen feet in diameter, no bigger than a large bedroom. We were eight people in the group, and for the next month that tent would house us all.

Not only did we all sleep inside it, but we also set up our science lab there, our kitchen where all the meals would be cooked, and shelves to store our gear. It would never have worked if everyone on the team had not had such a positive outlook toward the whole endeavor.

We were three Americans: David, Brant, and myself. Then there was Amaraa, the driver, and Moogi, the red-cheeked, tough-as-nails cook. Moogi couldn't have been more than seventeen years old, but she worked incredibly hard (while also yelling at us a lot). Also with us were two students from Mongolia's national university. Like most Mongolians, they were easy to work with. As I learned, the Soviet culture of respecting science and academic institutions was firmly rooted in the Mongolian mindset as well. The final member of our team, Fish Erdenebat, was key, since he was the only one among us who really knew how to fish for taimen.

In the beginning we didn't have a boat, so were limited to fishing from the riverbank. It was possible to catch taimen from the bank, but it meant we could not access many of the best spots. Naturally, we took our lead from Erdenebat, though we each had our own gear and soon began to develop our own techniques. David used a fly rod, with giant and gaudy homemade flies. Erdenebat used a large silver lure, called a spoon, while Brant and I fished with broken-back Rapalas, jointed fishing lures that are designed to wiggle like an injured fish. Initially we were not very successful, except for Erdenebat, who seemed to be able to find fish around every bend. Occasionally we would also go snorkeling. We could see fish that way, even if we had trouble differentiating taimen from their smaller cousin, the lenok. But snorkeling was a bit dicey, since we didn't have a boat and there was no one to retrieve us if we floated too far downstream.

Our access to fish improved when we partnered more closely with Sweetwater Travel, the fly-fishing outfit. Each morning, one researcher would join a Sweetwater guide and two paying clients in a small, flat-bottomed jet boat and head out on the river. The guide was the boss, responsible for driving and maintaining the boat; choosing the fishing spots; deciding how long we would fish and who would fish where; and coaching anglers on various techniques and how to get a hooked taimen into the net. It could be a difficult and thankless job, with a day's hard work in no way guaranteeing a catch.

But soon we had our hands on twenty to thirty taimen per week. Our research protocols were fairly straightforward: we measured and weighed each fish, took a photo and fin clip, and then inserted an external numbered tag below each fish's dorsal fin. The tags allowed us to identify the fish if it was recaptured, which in turn enabled us to measure growth and estimate population size and movement patterns.

The partnership with Sweetwater definitely kick-started our research, though it had its drawbacks. The company's priority was its paying clients, and those clients' priority was fishing, not watching us tag and measure fish. Eventually we would need more independence, both because we planned to start more time-consuming research (acoustic and radio tagging, rather than external tagging) and because Sweetwater operated for only six to eight weeks a year, whereas we planned to conduct fieldwork from April through October each year.

So, we bought a rubber raft. Holding four people, it was well made and very sturdy. With help from the Mongolian team members, David built a frame for it to attach oars made from scrap wood. The raft could be transported on top of the van, and sometimes Amaraa, the driver, would take us far upriver and drop us off. We estimated how long it would take us to get back to camp, sometimes planning for a long day trip. Other times we would bring tents and food and go on two-to-three-day excursions. Fishing from the raft was not perfect because there was no real way to position the boat or hold it in place at a good fishing hole, but it was peaceful, quietly drifting downriver with no one for miles, and more effective than shore fishing.

I loved life in our camp, even if it was pretty rough living. Water had to be brought from the river in large jugs for drinking and washing, though I remember drinking only tea (since the water had to be boiled) and not bathing much at all. The bathroom was a pit latrine dug on the periphery of the main camp and somewhat tricky to find in the dark.

Inside the ger, it was organized chaos. Everyone slept on a big rug at the back, with a wood-burning stove in the middle that sent the temperature inside yo-yoing as fresh logs were placed in it. All the meals consisted of mutton served in various forms: as part of a stew, with noodles, or fried into hoshuur, the traditional Mongolian hot pocket. With half an animal carcass hanging from the ceiling and walls insulated with felt, there was the constant and overpowering smell of wet sheep.

We also had to deal with the unpredictable weather. In the fall, most of the days were still sunny and warm. But then, out of nowhere, a blizzard could appear, as I found out on one of my first outings on the river. The sky was clear and blue, and because I was going out with Sweetwater clients to fish I was dressed in sandals, a T-shirt, and a pair of thin pants, which turned out to be a big mistake. Suddenly, dark clouds swept in across the valley, and it quickly got very cold and we were far from the camp. I had been wading in the river earlier and my pants were wet from the knees down. As it started sleeting, I had to admit to myself and the others in the boat that I was wet, cold, and embarrassingly unprepared.

Since I was a guest researcher on a boat with paying clients, I couldn't ask to be taken back to camp. Instead, the fishing guide got a fire started for me on the riverbank. I was able to warm up there while the guide and the clients continued to fish, then retrieved me at the end of the day.

As the weeks wore on, it got progressively colder, just as it had during our first trip to Mongolia, and soon it began to snow. But when it came time to leave, I was sad to go. Despite the hardships, the month we spent in Mongolia was so different from what I was used to in the Mekong that it reenergized me in a way I hadn't realized I needed. It was clear that we had started something special with the project and that this was going to be important work. As the plane lifted off the tarmac in UB, starting the long trip home, I couldn't wait to get back.

Although I grew up in the Arizona desert, with a family that didn't fish, I have many vivid early memories of fishing. These memories are random and varied and involve a diversity of places and people, including some of my relatives, like my uncle Steve in Reno, whom I saw only once a year but who always took me fishing, and my older cousin Grady, who lived in Sacramento and whom I visited most summers.

Grady and I would grab some old fishing poles and walk down to the local park, which had a small pond, and there we would catch sunfish after sunfish, using corn or cheese or salmon eggs for bait, until our bucket was full and we'd dump the fish back in the pond and go home.

In Tempe, my friends and I rode our bikes to Kiwanis Park, and there we'd catch small catfish from dirty ponds, marveling at the strange creatures before releasing them and seeing what else we could capture. At carnivals and fairs, I sought out games with goldfish as prizes, which my parents were none too happy to see me win because they didn't want another pet. As I got older, I began seeking out fish adventures on my own, in places like Lake Tahoe and Washington State. On a high school exchange program in France, I went ocean fishing with my host family, who caught a six-foot-long conger eel, which they hung in their garage and from which they spent the next week chopping off meal-size chunks to serve for family dinner.

In those early days I fished by happenstance and viewed it as an adventure. Fishing was a reason to explore and go somewhere new. Later, when I began to study fish, my relationship to fishing changed. As an undergraduate student working on the Colorado River, I was no longer fishing for a lark but with the specific purpose of catching fish that could be studied. In the Mekong, the fish I was researching were caught by other people. Fishing was at the heart of everything I was doing in my work, and yet it became secondary to the study of the actual fish, as aquatic animals, and the rivers where they occurred.

It was not until I came to Mongolia, and began returning there, that my appreciation for angling—that is, fishing with a rod and reel—deepened. Up to that point, the limited angling skills that I had acquired had been the result of sporadic attempts over the years to catch rainbow trout or the odd catfish. What I lacked in skill I made up for in enthusiasm. Things changed in Mongolia. There, I would fish all day for many weeks at a time, all the while being around people who loved to fish for the sake of fishing—and who were very good at it—and soon both my angling attitudes and abilities began to shift.

A lot of it had to do with the fish we were catching (and releasing): the taimen. The fact that it was the world's largest trout obviously made it a tantalizing catch. It was also a beautiful fish with a stunning coloration: a gray-green streamlined body with cross-shaped dark spots blending to an intense red color near its rear fin. But what made the taimen such an amazing game fish above all was its ferociousness, the way it would strike at the surface like a great white shark leaping after a seal.

Anyone who has ever caught a fish knows the rush of excitement that comes with feeling that tug on the line. Taimen take that sensation to a whole new level. The fish will absolutely explode on the fly or whatever it is attacking. So intense is that moment that first-time fishers will often instinctively jerk the line from the water and forget to set the hook. At times, the fish will go completely airborne on the strike, or, like a big tarpon, it will tail walk along the river's surface.

Anglers remember those strikes. I certainly remember the four-foot taimen that jumped out of the water into the scrub to take a fly that had been tangled by an errant cast; the five-footer that first jumped after Erdenebat's lure but missed, only to attack mine seconds later; and the two yard-long beauties that swirled around my fly one day, a whirling ball of silver only an arm's length behind the boat, before one of them pounced.

It was a fish that instilled a deep respect among anglers, not only because of its size and strength but also because strikes don't come often. When they do, it takes time and skill to land the fish. We also learned that taimen can live for fifty years or maybe even more, and we would catch the same big fish in the same deep pools year after year.

Trying to catch a taimen on a fly rod is no easy task. Like others who fish for taimen, I became extremely focused when I had one on the line. Landing a taimen required equal parts of luck, skill, and spatial awareness, because the taimen is so strong and fast. When hooked, it could dart downriver in an instant or rocket underneath the boat, as it pulled line off the reel with the rod bent into the water. With the movement of the boat, the flow of the river, and numerous people on the boat trying to help or get out of the way, there was a lot of quick motion and chaos.

Given that we were fishing with barbless hooks, and that most taimen, especially big ones, energetically pulled, shook, spun, and jumped, a "fish on" did not equate to a fish in hand. A strike resulted in a successful catch maybe only 30 percent of the time. These sequences left a permanent imprint in my mind, made all the more powerful because they were so rare. Sometimes we would go morning to dusk without any action, and then, on our last cast, we would finally experience the quick burst of adrenaline of the day's first strike.

That too was something I appreciated about fishing for the taimen: it brought me out on the water for long periods of time, allowing me to experience the water, observe how it changed and behaved, and in doing so form a meaningful connection to the river. I watched its eddies, riffles, rapids, and swirls, making mental notes of the river conditions and guessing where the next taimen might bite. I came to know every spot where we had landed a taimen—every bend, shallow section, drop-off, and pool. I saw the river change, from when it thawed in the spring and flooded in the summer, to the time when the water turned clear in the fall, as the surface filled with yellow needles of larch reacting to the winds and chills of the winter drawing near.

The more time I spent with anglers, the more my interest grew in the varied reasons people fish for sport. I can think of few activities whose practitioners are as dedicated to and passionate about as angling. Through the years I have fished with hundreds, maybe thousands of anglers—people who travel to other continents to fish and people who fish from their back porch—and their motivations are as diverse as the fish they catch.

There are the masters of the craft, or those who aspire to be. People for whom angling is an art form to be perfected. Most fly-fishing guides fit in this category. They are people who can make casting the entire width of the river look beautiful and effortless. There are naturalists, and to a large extent I include myself among them; those for whom fishing is a way to get immersed in nature. Learning about fish—the variations in color patterns and behaviors, or the environmental conditions that cue fish to spawn— teaches us larger lessons about the workings of nature.

Fishing becomes a chance to explore extraordinary landscapes, like the Zhupanova River in Kamchatka, in Russia's far east, where the headwaters flow through an immaculate valley framed by smoking volcanoes, snow-capped mountains, and ancient birch forests, and which is home to prolific populations of salmon, Dolly Varden, and some of the largest rainbow trout on the planet.

For many people, the allure of fishing is the simple diversion it provides. It allows someone to spend an afternoon with a nephew to teach him how to fish. For others, it is a chance to categorize and check things off, like for my friend Len Kouba, an angler with whom I crossed paths several times on my journey of chasing giants but who, sadly, passed away recently. An affable former university professor from Chicago, Len was a man with a quick smile and a story. Late in life, he devised an interesting bucket list for himself: he would catch as many different freshwater fish over one hundred pounds as he could. His list of species mirrored my own, from stingrays and sturgeon to catfish and gar, and over the years he traveled all continents in search of freshwater whoppers. The last time I spoke with him, he was closing in on a dozen catches. But of course, Len was not just out to check off fish on his list. He also loved exploring new rivers, being out in nature, and he wanted to see what adventure awaited around the next bend.

Anglers talk about fishing's enigmatic appeal, about something that is not explainable but deeply felt. There is undoubtedly something exciting, almost primal, about that moment when the fish takes your bait. You know the fish is there, somewhere down in the water, maybe deep underneath you. Often you can feel it, but you can't see it, you don't know how big it is, or how it might behave. You're interacting with a creature that inhabits the same planet as you do, and yet belongs to a world that is alien from your own. In that moment you are connected to a different dimension, and it is a feeling like no other.

Although I have come to appreciate all the facets of fishing, I have always approached it from a science and conservation perspective. I've tried to emphasize this in the television show: that I'm not a professional angler but a biologist and a nature lover. Ninety-nine percent of the time that I am fishing, I am fishing for research. I am rarely alone, but usually with other experts, whether they are scientists or local fishers. These are people with an intimate knowledge of the fish, often gained over decades, who can share that knowledge with me. Fishing helps me tell the stories of these large fish, and so it is a means to an end, not the end itself.

But I love talking about fishing with kids who watch the show. They remind me of myself as a young child, when I would get up early to watch the nature shows on TV. The kids that I meet are wide-eyed, curious, and optimistic, traits that will hopefully put them on a path to caring about fish and freshwater systems in the future. Despite all the discoveries and progress that have been made, it feels like we are still in the early stages of a true, global movement in which people care about rivers and fish, and for that to happen we will need these kids to take the helm.

There is one more story I want to share about anglers and their reasons for fishing, which speaks to that undefinable element of it. Early in my time in Mongolia, when I was working with Sweetwater Travel, there was a client coming in from New York to fish for taimen for a week. The client was an investment banker and clearly very wealthy, as he flew into the country on a private jet. He paid extra to fish alone (normally there were two clients per boat), and he didn't say a word while we were out on the river. It became immediately clear that he was an expert angler. At the top of each run, he'd jump out of the boat and into the river, casting hard and far until he had fished every inch of water that he could reach, and then he would climb back into the boat and sit silently while we traveled to the next spot.

On the last day, Charlie Conn, our guide, took us to one of the best fishing holes on the river, which we called Swimming Dog. Charlie positioned the boat perfectly and the banker immediately began to cast. We had been there for a while and were considering moving on when the guy suddenly got a strike. He expertly brought in the fish. It was the largest taimen I had ever seen, about five feet in length. I was beside myself with excitement, and so was Charlie. This was a once in a lifetime catch. As we pulled the fish to the side of the boat, we asked the angler if he wanted a photo of it. To our surprise, he seemed uninterested, and before we had even tagged and released the fish, he went back to fishing. The next day he flew back to New York. I never saw him again.

Documenting catches or the occurrence of large fish appears to be a practice as old as fishing itself. In Pha Taem National Park in eastern Thailand, cave paintings dating back more than three thousand years depict people next to enormous catfish. Visitors to the ancient temples of Angkor are greeted by bas relief carvings more than a millennium old detailing a menagerie of fish as large as crocodiles, from giant sheatfish to giant barbs.

Fast forward to the 1800s, when more specific records of giant fish, including credible measurements of lengths and weights, begin to regularly appear. A 3,249-pound beluga sturgeon was reportedly caught in the Volga River in Russia in 1827. The website for Guinness World Records lists it as the largest fish ever recorded in fresh water (though the beluga is not strictly a freshwater fish).

By the early 1900s, the popularity of sport fishing had been established and with it the demand for more accurate record-keeping, as anglers of various skill levels set out to break records by catching the largest fish. While some rules pertaining to fishing conduct had popped up at fishing clubs here and there, there was no universal code of fishing ethics or guidelines to steer anglers in their pursuits. This is what eventually led to the establishment, on June 7, 1939, of the International Game Fish Association (IGFA) at a meeting held in a cramped upstairs office of the American Museum of Natural History in New York City.

Today, the IGFA is headquartered in Dania Beach, Florida. Considered the world's governing body for sport fishing, it tracks records based on size of species, fishing gear, line class, and habitat (salt water vs. fresh water) for male, female, and junior anglers. In short, it is the authority on fishing records, and sport fishers are careful to follow the organization's stringent rules in order to achieve the honor of being listed in the IGFA's annual *World Record Game Fishes*.

For a project like mine, which, on one level, is a record-seeking endeavor, the IGFA is a useful resource, though not as helpful as one might think. Although IGFA records can be trusted, they are far from comprehensive. There are no records prior to the organization's founding; no records from commercial harvests; and no records from some of the more remote regions of the world. The process of certifying a record with the IGFA is so complex, with a plethora of how-to videos, that most catches of big fish simply don't meet its requirements. For example, a quick look at IGFA records for Mekong species such as the giant catfish (260 pounds) and giant barb (143 pounds) reveals that IGFA record fish often weigh significantly less than catches verified away from the IGFA process and rules.

There are problems with other record databases as well. Although Fish-Base, operated and curated by scientists, is considered by most fish biologists to be the gold standard for all information about fish, it is far from bullet proof. While FishBase provides referenced information about a fish's ecology, distribution, and maximum size, meaning it's possible to see where the information was first published, those references are sometimes questionable or outdated. To make matters worse, scientific journals may repeat the erroneous information, because it has been reported by a reputable scientific source. According to FishBase, the Murray cod, one of Australia's largest freshwater fishes, grows up to 250 pounds, but the numerous Australian biologists I've spoken to say it has never grown that large.

One useful website that appeared around the same time I started my search was www.fishing-worldrecords.com, which documents rod-and-reel catches around the world. The site includes questionable records of enormous catches that could not possibly be verified, but it is updated regularly and is easily searchable. Most helpful, it has a "what's new" section in which photos are posted of recent catches with a date and location attached. I have found that it is usually possible to tell by a photo, if it is relatively recent, whether or not a record is credible. Most of the photos published on this site appear legitimate, and sometimes they have provided me with a starting point for my own research or field visits.

It's usually also possible to ascertain a report's credibility by talking to the person who caught the fish. Through my work I have had a chance to meet many of the people who hold fishing records. On a recent trip to open the National Geographic Monster Fish exhibit at Exploration Place in Wichita, Kansas, I met local amateur fisher Grant Rader, who was fishing on Keystone Lake in Oklahoma for his eighteenth birthday when he hooked into a 164-pound American paddlefish. Grant had been out with a professional fishing guide, who photographed and weighed the fish, so the record was definitely authentic. (Keystone Lake also happens to be the source of the previous catch record for the American paddlefish.)

My favorite moment from the museum opening came when Grant saw the exhibit's paddlefish sculpture, modeled to the maximum reported size of the species, and called it "small." Indeed he had a point: I had always viewed the exhibit's replica paddlefish as impossibly large, but when I saw a photo of Grant's catch in the local paper, I had to accept it was bigger.

Since the vast majority of fish are caught for commercial reasons, it makes sense that a record-breaking fish would be caught by commercial fishers, who also have more tools at their disposal—large seines, gill nets, traps, trawls, and trotlines—to catch big fish than rod-and-reel anglers do. But commercial fishers fish for their livelihoods, not to break records, and they may not even weigh or have an interest in reporting a potential record-breaking catch. The Thai fishers who caught the 646-pound giant catfish in 2005 came from a community that's been fishing for food and income for decades, and where people had undoubtedly caught some very large catfish before. The record-breaking fish was not initially viewed as anything too out of the ordinary. To the fishers, it was simply a fish that was a bit bigger than the last.

One question concerning records, which arose early in my quest for answers, was if a potential record would count if the fish had been caught in an artificial setting. The question had gained relevance because of the proliferation of artificial fishing ponds in Thailand. These were places where recreational anglers were charged for the opportunity to fish for some of the world's largest species, which had been collected from freshwater systems near and far and stocked in the ponds.

One such place was Gillhams Fishing Resort outside of the city of Krabi in southern Thailand, which I first visited in 2010. Nestled in between green, lush mountains, the resort was run by a good-natured Englishman named Stuart Gillham, who had sold a successful scaffolding business back in the UK to come to Thailand to start his dream business. The resort consisted of a dozen or so bungalows that had been built around a twelve-acre lake that had been stocked with more than fifty species of fish, including giants such as arapaima, Mekong giant catfish, and Siamese carp. The latter species was a main draw for the British visitors, who made up most of the resort clientele. For some reason Brits are obsessed with carp fishing.

As it turns out, Gillhams is a pretty good place to be a monster fish. Free of harvest, pollution, and predators, the giant fish are allowed to grow in a controlled and engineered environment where they thrive. And grow they do, often outpacing sizes seen among their megafish cousins in the wild. Several IGFA rod-and-reel world records have been broken at Gillhams, including for the Mekong giant catfish (260 pounds) and giant Siamese carp (134 pounds), and other, bigger fish have been caught there that did not adhere to the strict IGFA rules. The lake is home to arapaimas said to weigh over 500 pounds.

But what if an arapaima from Gillhams grows larger than the record arapaima from the wild, or even the record-breaking Mekong giant catfish? Should the arapaima be considered the world's largest freshwater fish? Many people would say no, because fish at Gillhams benefit from supplemental food and other conditions that could conceivably create an artificially large, or fat, fish. I can certainly see their point, but for me it is not a question of where the fish was caught or under what circumstances. My goal, as it pertains to the record, is to find out how big these fish can grow, regardless of their habitat.

I do believe that fishing at Gillhams and similarly well-operated fishing resorts can provide conservation benefits. Instead of removing from the wild large, adult fish, which are the most fecund (fertile) and thus most important for species survival, recreational anglers are able to catch and release captive fish, leaving wild stocks unperturbed. This is particularly important since fishes targeted for records are disproportionately represented among the most vulnerable fish species in the world. A 2014 study published in the journal *Marine Policy* showed that eighty-five species for which IGFA records have been issued are classified by the IUCN as threatened.

It was clear that records played an important role for the project. Only through a verified record, or credible documentation, could I hope to answer the question that I had set out to ask. But initially I overestimated the number of records and data there would be to deliver that answer. I had not appreciated the ambiguities associated with records, and soon I came to understand that in my search, records, along with stories from fishers and commercial catch statistics, provided a starting point for my work, not the end. My goal has never been to break angling records.

What was clear was that almost all of the megafishes were getting rarer, and I realized that finding a record breaker would probably only get harder. With increased harvesting in many places around the world, the giant fish in the wild did not have time to grow, and their average size kept shrinking. The six-hundred-pound giant pangasius and giant barbs that Hugh Smith had written about in *The Fresh-Water Fishes of Siam, or Thailand* in 1945 seemed to be long gone. Specimens of these species that were still found tended to be all juveniles weighing no more than five pounds. In this new world, as in the wake of the Devonian extinction some 360 million years ago, smaller fish, which can reproduce at a younger age, were becoming evolutionary winners.

It seemed to me that, as the world's giant fishes disappeared, any and all of the records and acknowledgments of their historic size would take on greater importance. Which is why I found it so disturbing when, one day, I went to look up the Guinness World Records entry about the Mekong giant catfish being the world's largest freshwater fish and found that, for some reason, it had been deleted, and with it one small but crucial piece of confirmation that this extraordinary creature had once existed in that gargantuan form.

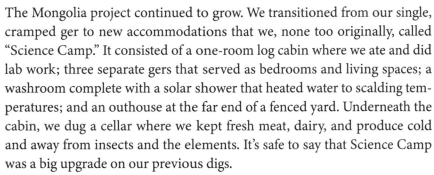

The Mongolia project continued to grow. We transitioned from our single, cramped ger to new accommodations that we, none too originally, called "Science Camp." It consisted of a one-room log cabin where we ate and did lab work; three separate gers that served as bedrooms and living spaces; a washroom complete with a solar shower that heated water to scalding temperatures; and an outhouse at the far end of a fenced yard. Underneath the cabin, we dug a cellar where we kept fresh meat, dairy, and produce cold and away from insects and the elements. It's safe to say that Science Camp was a big upgrade on our previous digs.

The new camp was situated farther from the river than our old one, three hundred yards back from the water's edge and a few meters above it. It was inconvenient to not be closer to the river, but we picked the spot, in an empty field, because of the flooding that sometimes occurred in the summer from violent rains swelling the Uur River and inundating the lower portions of the valley. It turned out to be a wise decision.

The US-based team members—David, Sudeep, Jake, Brant, and I—rotated in and out of the camp from spring to fall, while the Mongolians (students and others) did the same, coming in from UB or the town of Erdenebulgan. One person who was from the area, and who became an instrumental team member from early on, was a young man named Ganzorig. He came from a small settlement called Teshig, which was located downstream from our study area. Ganzorig told us stories about growing up there and going fishing with his family members at night with lights and spearing taimen. At one point, they came across a taimen that seemed too big to spear, so they decided to shoot the fish with a gun instead.

Ganzorig was seventeen when he joined us and was starting his university studies. Over the following years he earned a bachelor's degree in biology and a master's degree in fisheries management. Although he had never held a fly-fishing rod before he came to work with us, he became a skilled fly fisherman. Being from the area, he knew the river better than anyone else and was a huge asset to the project.

The camp received a constant flow of visiting scientists, students, writers, and filmmakers. These were people who had heard about the project and wanted to spend a week, month, or even a whole season with us in the Mongolian wilderness. Through it all, the vibe in the camp remained extremely positive. Everyone was energized by the work, and within our team we figured out routines that suited each person's personality. There were main undertakings, like the tagging of fish, and side projects, like developing a photo identification system based on the spot patterns of the fish. No ideas were too outlandish as long as we thought it might work. At one point we built a huge wooden tower, which we placed on our boat. From there, we thought we could conduct a fish census, but it didn't work very well.

When we weren't working on the project, we went for hikes or horseback rides through the birch-brimmed countryside. We also developed a strong connection to the surrounding communities, who were all very welcoming of us. Since some of the Mongolian members of our team hailed from the area, we regularly got invited to events and celebrations. One such outing landed us in a spot of trouble. We were on our way to a wedding, with David driving the van, when suddenly he lost control on an icy hill and crashed into a birch tree. The crash resulted in David losing a front tooth from hitting the steering wheel, but it could have gone much worse.

It was hardly the only mishap we experienced. There were other car and boat crashes, broken-down vehicles, canceled flights, bee stings, food poisonings, sunburns, heat strokes, and so many wet and cold boat rides that I lost count after the first year. Then, one July, the area experienced a once-in-a-century flood. I wasn't there at the time, but David was. As the intense rains filled the Uur, the river rose to the doorstep of Science Camp. Fortunately it stopped there, and our camp did not suffer any damage. Many of the locals were not so fortunate, ending up losing both houses and livestock. David spent two frantic days driving our project boat around to save people's drowning sheep. He later received an award from the Mongolian government in recognition of his actions.

It was all wild and we loved it. But all good things must come to an end. Five years had passed, and in 2009 the project was wrapping up. It was time to take stock of what we had accomplished, most notably what we had learned about taimen ecology. We had determined that there were approximately two thousand "catchable"-size taimen in our study area, or about twenty fish per kilometer. The smallest fish we caught measured just twelve inches, while the largest was sixty inches (five feet) and weighed almost seventy pounds. We had not found the six-foot, two-hundred-pound fish that we had seen in historical accounts, but the largest adults we caught were still leviathans. They were as old as I was at the time, and capable of taking very large prey. One fish that we recovered had choked on an eight-pound muskrat.

We had established that the taimen made seasonal migrations, especially in the spring and fall, and seemed to overwinter in groups in deep pools. A few of the fish we tracked had migrated over one hundred kilometers, but most kept their travels to around twenty kilometers, with individual fish returning to the same spots year after year. We had figured out what the fish ate and when, how fast they grew, and when and where they spawned. This was information essential for management guidance.

Our conclusion was that catch-and-release fishing for taimen was sustainable, as long as the accidental mortality rate was kept to around 2 percent. It was clear that intensive harvesting of the fish would quickly remove the largest taimen from the system and ultimately doom the population. Taimen needed clean, cool, flowing water and gravel substrate to spawn; the muddy, stagnant waters that we had seen in the heavily mined areas would not work. Neither would rivers that were dammed.

While the science component of the project was an unqualified success, it was different with the fishing concession model that had got it all started. The model was ultimately not adopted by the Mongolian authorities. Letting predominantly foreign anglers fish the Uur and Eg Rivers while most Mongolians were not allowed to do so was a difficult concept to sell to the public, especially in a country struggling to shed its foreign dependence. (It's worth noting that a similar management regime for the taimen has since been established in the Tugur watershed in Russia, though on privately owned land, which doesn't really exist in rural Mongolia.)

Around the time the project ended, a large group of American and Russian researchers, including myself, went on a tour around Mongolia to study the general environmental impact of mining in the country. The scramble for Mongolia's natural resources had intensified and grown more complex, pitting powerful foreign multinationals against domestic business interests and small-scale mining operations. With the Mongolian government unable, or unwilling, to prioritize environmental protection, it was increasingly left up to nonprofit groups and local grassroots organizations to defend nature.

One prominent leader in this environmental movement, and someone who led us on our study tour, was Tsetsegee Munkhbayar. A charismatic and energetic man, Munkhbayar had spent his childhood herding yaks on the banks of the Ongi River in central Mongolia. In the early 1990s, he had witnessed the river, an essential life source to sixty thousand people, begin to shrink and grow contaminated as a result of unregulated hydraulic mining for gold and other minerals. Desperate for drinking water, residents had no choice but to dig wells, but at that point the groundwater had become so contaminated that many local children ended up suffering serious liver damage. Munkhbayar's own son was taken ill, and his mother lost her life.

By the turn of the millennium, the river that had coursed through his village for centuries had completely vanished, along with more than thirty tributaries. That's when Munkhbayar decided to fight back. He got elected chairman of his local citizens council, which was highly unusual for a herdsman, and cofounded the Ongi River Movement, bringing water issues to the attention of government officials, initially at the provincial level and later on nationally. Tireless, he held press conferences, organized town hall meetings, and mobilized protest marches involving thousands of people, unheard of in Mongolia.

Remarkably, his efforts appear to have paid off. The movement convinced government officials to strengthen mining regulations, pass new legislation, and endorse environmental restoration work. As a result, thirty-five of the thirty-seven mining operations in the Ongi River basin were said to have stopped their destructive practices, with the worst offender shut down completely. Soon, the river began to trickle back. In recognition of his work, Munkhbayar was awarded the prestigious Goldman Environmental Prize in 2007. National Geographic named him an Emerging Explorer. By the time we met with him, he was in the process of forming the Mongolian Nature Protection Coalition, an organization uniting eleven river movements in the country.

But with more than half of the nation's land granted to mining, there was clearly a lot of work left to do. Everywhere we traveled, we were faced with scenes of environmental devastation: entire mountainsides blown off, lakes emptied, steppes turned to desert. Democracy, Munkhbayar said, had brought a lot of good things to Mongolia, like freedom and openness to the world. But it had also brought foreign mining companies that saw the country as a source of mineral riches they could exploit. As he was speaking, he was standing in a dried-out tributary of the Ongi, near where he had grown up, and where the water had once reached up to his chest.

This time I left Mongolia conflicted. Munkhbayar and some of the other people we had met on the trip had been a great source of inspiration. But the forces they were up against were so formidable and the economic interests so strong that the challenge to bring about systemic and lasting change seemed insurmountable.

Then, a few years later, I got news that Munkhbayar had been arrested and condemned to a twenty-one-year prison sentence for opening fire in a public space. Apparently, the incident had occurred during a demonstration in front of a Mongolian parliament building, where a rifle had been discharged. Although the firing of the rifle was said to have been accidental, and no one had been hurt, Munkhbayar and several of his fellow protesters had been charged with "acts of terrorism."

It was a charge that many people saw as politically motivated. Munkhbayar and others had been protesting government attempts to weaken the "Law to prohibit mineral exploration and mining at headwaters of rivers, protected zones of water reservoirs, and forested areas," or, as it was known informally, "the law with the long name." Thankfully, Munkhbayar was pardoned two years later.

We all wanted to continue working in Mongolia. The country was in political flux and clearly under a lot of environmental pressure. Protecting the taimen, and ensuring its future survival in Mongolia, was something we all felt very passionate about. We had produced the best science available on the ecology and conservation of the species, but there was still a lot of work to do. It would be a shame if we could not continue, especially as the Taimen Conservation Fund (TCF) remained active. But without renewed funding, there was little we could do, and so we were forced to go our separate ways. I shifted my focus to research in other regions and to filming *Monster Fish*. From 2009 to 2015, I filmed over twenty episodes, each one focusing on a different location and different fish.

But a few years after the end of the program, a postdoc at the University of Wisconsin named Olaf Jensen received funding from the National Science Foundation to study aquatic food webs in Mongolia. He teamed up with the TCF and picked up where we had left off on the Eg and Uur Rivers, and also on Lake Hovsgol, one of Mongolia's largest alpine lakes. After bringing students from the United States to do some summer projects, Olaf eventually got additional money from the TCF to investigate the likely impacts of a dam that had been proposed for the lower Eg, just above its confluence with the Selenge.

This was an issue that was right up my alley; I had started my career by studying the potential impact of dams on migratory fish. As it happened, Sudeep had also received support from the NSF to study large-scale patterns in aquatic ecosystems in the United States and Mongolia. Meanwhile, I was getting financial support from National Geographic that I could use for travel and research on taimen. It was all coming together. So in 2014 the whole team joined up with our colleagues in Mongolia again, this time to assess the risks presented to the taimen by the planned dam.

From everything we've learned about dams, it's clear they almost always change downstream environments in ways that are detrimental to native fish and other biodiversity. Yet the ecological impact in this situation was, as is always the case, hardly the only consideration. Some of the dam advocates were playing up the "green" potential of the dam compared to the dirty coal-producing energy that Mongolia would otherwise be forced to use. Meanwhile, Russia was opposed to the dam, arguing that it would negatively impact Lake Baikal downstream.

Our work on the river around the proposed dam site mirrored the work that we had done a decade earlier and a hundred miles upstream. We caught and tagged fish to estimate population numbers, examined growth trends, and looked at movement patterns. We were particularly interested to learn if there was movement between the Eg River and the Selenge, because the dam was planned for a site just a few miles up from the confluence.

Life along the Eg and Uur Rivers looked much like it had when I first arrived there. But beyond the mountains, and just downstream on the Selenge, you could sense the changes that the country was going through, and it was just a matter of time before those changes could be felt at our study site. In truth, it was already happening. Every year that I returned to Mongolia, traveling north, a new stretch of highway seemed to have replaced what was once dirt road, attracting an ever-steadier stream of Land Cruisers and Priuses. Tall wooden fences appeared out of nowhere to mark personal properties where once the land was wide and open. And more and more people came to fish, as cultural views of the taimen changed.

This was not necessarily a bad thing. Armed with the knowledge showing that taimen populations could not handle extractive fishing, lawmakers passed legislation requiring that all fishing for taimen had to be done on a catch-and-release basis. There was growing interest in fly-fishing in the country, with fly-fishing clubs and lessons increasingly available for budding enthusiasts. I was not so naive to think that challenges did not lie ahead. Transparency and enforcement were going to be major issues to work out. But I also knew that anglers can be powerful advocates for fish conservation, and that there are many other places where the kind of approach that

Mongolia is taking has been successful, from trout fishing in Montana to salmon fishing in Alaska. Maybe Mongolia can follow suit?

My last trip to Mongolia before COVID-19 stopped international travel in its tracks was in October 2019. In UB, where cars had been banned one day a week because of the pollution, the political discourse had grown heated. Many observers lamented a turn toward autocracy in the country; a proposed law was expected to marginalize smaller parties and remove the rights of Mongolian expatriates to vote.

Arriving at Science Camp felt like coming home. Cows and sheep grazed in the pastures adjacent to our gers, just as they had when we got started fifteen years earlier. Most of the Mongolians who were there at the beginning were still there. Guides who were young back then were now married with children, and parents had become grandparents. The river, meanwhile, was flowing low and clear, and on our first afternoon we went fishing and caught a nice taimen from "home pool," an area of slow-moving, deep water about two hundred meters upstream from camp.

One day we encountered two young Mongolian herders, and Ganzorig noticed that they had a taimen wrapped in a towel on the riverbank, apparently in an attempt to hide it. Ganzorig went to speak with them. Somehow, the fish was still alive, and so Ganzorig took it and released it back into the river. Incredibly, it swam off.

Soon after that, we encountered a whole family floating down the river in different rafts piled high with gear. I noticed pots and pans and old tents and fishing rods. They were moving through the core section of our old study area, and as we passed them on the river they all smiled and waved. As it turned out, they had come up from UB for a weeklong family vacation. It looked like a proper adventure.

When my own week had passed, another researcher asked Ganzorig if he would be interested in floating the river from the confluence of the Uur and the Eg to the confluence of the Eg and the Selenge. It would take a week. By this time, temperatures had dropped well below zero at night, and it was so cold during the day that ice formed on our fishing lines. But that did not deter Ganzorig. He was always up for a trip down the river. You never knew, he seemed to reason, when you might come across a six-foot taimen.

Close Calls with the God of Darkness

Many of the divine beings in Buddhism derive from Hindu mythology, and so it is with Rahu, the "god of darkness," whose origins are laid out in the Vedas, Hinduism's oldest scriptures. The story begins a long time ago in a spiritual galaxy where constant wars raged between benevolent gods, called devas, and demons known as asuras, one of whom was named Swarbhanu.

At one point, the gods all came together to partake of a divine nectar called amrit, which promised immortality for those drinking it. During this ritual, Swarbhanu figured out that the devas were planning to deny the elixir to the asuras, and so he disguised himself as a deva to sneak a sip.

When the chief deva, Lord Vishnu, noticed what was going on, he reacted in a rage by hurling a discus at Swarbhanu, cutting his head off. Only it was too late. Swarbhanu had already consumed a few drops of the elixir, instantly making his head immortal. It was then decided by Lord Brahma, the creator god, that the head would have a snake's body attached to it, creating a new god named Rahu.

Rahu's star instantly shot to prominence, and he was named the eighth planet in Hindu astrology. Another deity called Ketu, whom Brahma had created from the leftover head of the snake and the body of Swarbhanu, was named the ninth planet. Together, the two planets came to represent the points in space where the orbit of the moon crosses the path of the sun. It was believed that solar and lunar eclipses occurred as Rahu and Ketu swallowed the sun and the moon, respectively.

It's unclear when Rahu was brought into the Buddhist pantheon, but we know he makes the leap from Hinduism without Ketu, who doesn't figure in

Buddhism at all. In Buddhist tradition, it is Rahu, the one god of darkness, who causes both solar and lunar eclipses, which are believed to end when the sun or the moon passes the opening of Rahu's neck.

As one of the krodhadevatas, or terror-inspiring gods, Rahu represents uncleanliness, falsehoods, and harsh speech. But he is also associated with material manifestation and worldly desire. Abrupt changes in fame and prosperity are linked to him. In Thailand today, seeking Rahu's blessing to repel bad luck is a common practice. He can strengthen one's power and convert an enemy into a friend.

He is, in other words, a complicated and mysterious being, operating in the shadows of the murky netherworld, much like a river creature that has been given his name, the one Thais call Pla Gra Ben Rahu.

When I first arrived in Southeast Asia a quarter of a century ago, I had never heard of the giant freshwater stingray. Like most people, I thought of sting-rays as marine creatures, not fish that swam in rivers and lakes. I had seen rays in the wild, but only in the ocean. Once, on a family vacation to Hawaii when I was about twelve years old, I took a walk along a seaside cliff after dark and came to a spot where the soft evening lights illuminated the water.

There, I could see dark shapes gliding through the water—huge manta rays coming into the shallows to feed. They were massive and otherworldly, and the sight of them was a highlight of my trip. But I did not encoun-ter or think about rays again for many years. Rays seemed to be strange outcasts of the sea, the ocean's shadow creatures that were not part of my freshwater world.

In fact, all rays are outcasts, at least in terms of study and conservation. Elusive and mysterious, they may be among the most poorly understood vertebrates on Earth. Just pinning down their numbers is tricky. There are an estimated 630 species of stingrays and skates in the world. (The differ-ence between rays and skates, broadly, is that rays [at least most marine species] are kite-shaped and have streamlined tails and venomous barbs, while skates tend to be roundish or triangular in shape, with heavier, fleshier tails, and have no stingers.)

The exact number of species, however, seems to be in constant flux, going up—as new species are discovered, which appears to be happening with remarkable regularity—or down, as happens when a lightly studied spe-cies thought to be distinct actually turns out to be the same as another one.

They are also a diverse group of fish. There are electric rays, butterfly rays, round rays, manta rays (a type of stingray without a stinger), and rays that don't look or even sound like they're rays, like guitarfish and sawfish. Belonging to the class of cartilaginous fish known as Chondrichthyes, rays and skates are closely related to sharks; so close, in fact, that they are easy to confuse. The angel shark, for example, is a shark, but it has a raylike body and looks more like a ray than a shark. The sawfish, on the other hand, is classified as a ray, but, with the exception of its elongated snout, it looks much more like a shark.

The vast majority of rays are indeed marine animals, with most found along the bottom of the shallow and coastal waters of tropical and temperate seas. There are also rays, such as the giant manta ray, which can grow up to thirty feet wide, that swim in the open ocean. However, there are about thirty-five species of rays that live in fresh water, mostly in rivers in South America, but also in Southeast Asia and Australia. It was one of those species, the giant freshwater stingray, that I had first seen with Niek van Zalinge in Prey Veng, Cambodia, in 2002. From that moment, I was pulled into the mystery of the giant stingray; my perception of the natural order of things had been reset. The stingray was so different from other freshwater giants. Its peculiar form suggested that the biggest fish didn't even have to look like fish. The rays broke the rules, though by then I had already begun to wonder: what were the rules governing the size of these fish?

A photo I took of the stingray was published online, and some people reacted with horror. They thought the ray looked terrifying. Perhaps it had something to do with the colors: the mud-brown, camouflaged body of the stingray against the chocolate-brown waters of the Mekong. It blended into the muddy waters so completely that even in an inch of water it would be invisible. Maybe it was the idea of stepping into that water and not being able to see this alien creature with its toxic stinger, only have it brush up against your legs. It didn't matter if I explained that giant freshwater stingrays, like other rays, were not aggressive creatures. There was something about stingrays that provoked irrational fears.

My research focus at the time was on the Mekong giant catfish, but sometime after my first encounter with the giant freshwater stingray I began a study with the Cambodian Department of Fisheries to assess populations of other large fish in the Mekong River, including the population of stingrays. Apart from the work I had done on the Tonle Sap River, hardly any population data existed on the largest species of fish in Cambodia, and basically none on the giant stingray. I wanted to know how abundant the rays were and how their populations compared to those of the past. I also wanted to know: How big did they really grow?

So I began traveling along the Mekong River, visiting local fish markets, just as I had done in Thailand years earlier. One of my first stops was Kampong Cham, a sleepy provincial capital about eighty miles up the river from Phnom Penh. It had a sizable Chinese population, which was explained by a curious legend. As the story went, a Cambodian boy had once been swallowed by a fish, which then swam all the way to China. When fishermen there caught the fish, they sliced it open, revealing the child, alive. The Chinese emperor decided to raise the boy as his own. Years later, the grown prince returned with ships full of Chinese sailors to populate the land that became known as Kampong Cham.

The market in town was buzzing and overflowing with fish, but I didn't find a lot of large species there. It was the same thing when I traveled farther north, visiting the markets in Kratie and Stung Treng, close to the Lao border. Occasionally, I saw chunks of freshwater stingray for sale, but never any whole rays. The pieces of meat displayed in the markets meant there were stingrays in the river. To learn more, I needed to find the people who were catching them.

That turned out to be more difficult than I had anticipated. The vast majority of fishers I met in Cambodia had never fished for stingrays. The rays were extremely tough to catch, and, because of their unwieldy shapes, they could not be caught with nets. Fishers had to use special equipment and expensive fish bait. Besides, the rays' rubbery meat was not considered high-quality food. For most of the subsistence fishers along the Mekong, fishing for stingray was just not worth it.

But there were those few willing to take on the challenge, and, with time, I began finding them—hardened men like Sok Long, who lived with his family in a tranquil village outside Kratie. A sinewy man with a kind face, Long was of Vietnamese background, and his father had also been a stingray fisher. Now Sok Long was teaching his own fifteen-year-old son how to catch the giants.

The technique he used was the same other stingray fishers used. Long would go out on his boat in the evening and set a trotline, a thick rope that could run up to a thousand feet long and which had a large rock tied at the end of it, acting as a weight on the river bottom. From the main line, at intervals of maybe sixty feet, branch lines made from heavy twine extended down with hooks at the end of them. On those hooks, Long put whole fish as bait. A float barrel tied to the trotline marked its location, and if it was moving when Long returned in the morning it meant a ray had been caught.

But hooking the ray was the easy part. Bringing it in was a whole different matter. Long and other stingray fishers told me stories about their long-tail boats being dragged by hooked rays for five, six, seven hours—for miles up and down the river. Lines would snap and gear would break. Even if the fisher managed to bring the giant stingray to the surface and hook into its flesh with the use of a gaff—a long stick with a crude hook at the end of it—the ray was often too heavy to load into the rickety boat, leaving him with no option but to drag the massive creature along with the whole trotline assembly behind the boat and onto shore.

Sok Long invited me into his house, built on stilts close to the river. Sitting on the floor, I interviewed him about catches he had made and how they compared to catches of the past. He showed me the old spines of rays that he had kept as souvenirs. He took me out on his boat on the Mekong, stopping at a sandbar in the middle of the river where he used his gaff to draw a picture in the sand of the largest stingray he had ever caught. The drawing was as big as a car.

Yet we never caught any rays during the time that I spent with Sok Long. The more I kept hearing stories about giant catches, the more elusive they became. I knew the stingrays existed. I had seen a giant myself, and the stories were too plentiful to dismiss. I had even found a peer-reviewed article by Ian Baird, which stated that fishers in northern Cambodia had caught stingrays weighing more than five hundred kilos, over a thousand pounds. But no one seemed to have talked to anyone who'd actually weighed a fish. There were so many unanswered questions about the giant freshwater stingray, beginning with its very identity.

———〜〜〜———

Modern taxonomy—the branch of science concerned with the classification of organisms—was invented by the Swedish botanist Carl Linnaeus, whose book *Systema Naturae* (or at least the tenth edition of it, published in 1758) laid out a system for arranging all living things according to structure and characteristics and placing them in categories of phylum, class, order, family, genus, and species.

Although scientists have been using this system for over 250 years, we haven't come anywhere close to classifying all living things on our planet. A study published in *Nature* in 2011 predicted there are about 8.7 million

Fish carvings on the walls of Angkor Wat illustrate the importance of fish to people living in the Mekong region over a thousand years ago. Many of the carvings depict species that still occur in Cambodia today and can be seen in catches from the Tonle Sap Lake, just a few miles from the temples. Photo by Zeb Hogan.

One of my first colleagues in Cambodia was Mr. Thach Phanara, who works for the Cambodian Fisheries Administration. In 2005 Phanara and I tagged approximately five thousand migratory fish moving out of the Tonle Sap River and into the Mekong. Photo by Zeb Hogan.

This giant catfish, at the time the world's largest recorded freshwater fish, was captured on May 1, 2005, by a group of Thai fishermen on the Mekong River in the Golden Triangle region on the border of Thailand and the Lao People's Democratic Republic. Photo by Suthep Kritsanavarin.

In the early 2000s, I worked with colleagues from the Cambodian Fisheries Administration, with support from the Mekong River Commission, to tag and release approximately twenty Mekong giant catfish weighing between two hundred and six hundred pounds. Photo by Zeb Hogan.

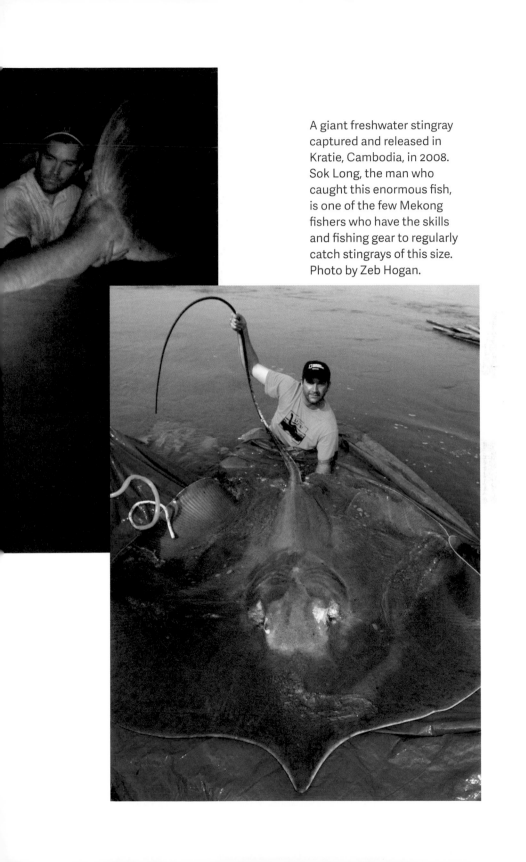

A giant freshwater stingray captured and released in Kratie, Cambodia, in 2008. Sok Long, the man who caught this enormous fish, is one of the few Mekong fishers who have the skills and fishing gear to regularly catch stingrays of this size. Photo by Zeb Hogan.

The alligator gar is one of North America's largest freshwater fish, growing to more than three hundred pounds. This gar was caught and released on the Trinity River in Texas. The Trinity is home to healthy, well-managed populations of the gar. In Texas the species is not commercially fished, and recreational fishing regulations include catch limits and closed seasons to protect the gar from overexploitation. Photo by Brant Allen.

The rivers of the Himalayan region of India, many of which flow clear in the dry season, are one of the best places to see and film goonch. We weren't able to snorkel or film in Jim Corbett National Park, where we saw goonch and crocodiles side by side in deep, rocky pools. But we did swim in the Ramganga River adjacent to the park, where local fishers caught this beautiful fish and let us swim with it for the tag and release. The coloration and speckling pattern of the goonch, together with its fleshy barbels and overall alien appearance, make it one of my favorite fish. Photo by Rob Taylor.

Northern Australia is home to one of the last healthy populations of freshwater sawfish. Sawfish live most of their lives in coastal water, but mature females give birth in estuaries, and young sawfish move into fresh water until them- selves reach maturity. Young sawfish were common catches in our short-set gill nets, and we would quickly remove them for tag and release. This research, led by Dr. Dave Morgan, was nerve-wracking because we had to be on constant lookout for salt- water crocs in water so murky we couldn't see what might be lurking nearby. Photo by Zeb Hogan.

The Daly River in Australia reminded me of tributaries of the Colorado River that I surveyed as an undergraduate at the University of Arizona: clear, fast-flowing creeks surrounded by red rock and characterized by riffles, pools, and small waterfalls. Unlike the streams of Arizona, the Daly River was also home to several species of large aquatic predators, two of which are pictured here: a bull shark and a freshwater whipray. Photo by Brant Allen.

Life in the field is full of challenges—it's often hot, wet, and dirty. All three were the case when we visited northern Australia to film a *Monster Fish* episode about the freshwater whipray. When a tropical cyclone hit soon after we started filming, we were deluged with wind and rain. On top of those challenges, we also couldn't catch a whipray because bull sharks—numerous and aggressive—kept taking our bait. Photo by Zeb Hogan.

At the beginning of our project in Mongolia, our accommodations were very basic: one Mongolian ger (also called a yurt) where we slept, worked, and cooked. We had no shower and used a neighboring herder's pit toilet. As the project grew, we built a log cabin, outhouse, and several more gers. Photo by Zeb Hogan.

My favorite memory from our first trip to Mongolia was of going up a small tributary of the Yeroo River near Khan Khentii National Park, the birthplace of Genghis Khan. Knowing that we would need to ford the river several times, we rode horseback in our wet suits, which made the ride more comfortable and the transition between riding and snorkeling seamless. A few times the river was so deep that the horse would swim, and we were able to stay with them the entire ride. Photo by Brant Allen.

In 2002, the University of California, Davis, the Cambodian Fisheries Administration, and the Mekong River Commission started a partnership, the Mekong Fish Conservation Project. The project, funded by National Geographic, supported several activities, most important the purchase and release of the critically endangered giant catfish and giant barb, including this release of a giant barb into the Tonle Sap River. Photo by Zeb Hogan.

Gathering data on the world's largest freshwater fish would be impossible without partnerships with local scientists and fishers. In this photo, I shake hands with Dr. Wei Qiwei after discussing the status and future of Chinese paddlefish. Dr. Wei was a coauthor of the 2020 study that announced the extinction of the Chinese paddlefish. The American paddlefish has so far avoided the Chinese paddlefish's fate. Photo by Stefan Lovgren.

The wels catfish has been introduced
into many rivers in western Europe,
where populations have flourished
and it has come to dominate native
aquatic biodiversity. Given the fish's
large mouth, enormous body size and
growth potential, and its proclivity to
eat almost anything, it's a formidable
predator. While it's not dangerous to
humans, the wels catfish poses a real
threat to native migratory fish such as
eels, salmon, and lamprey. Photo by
Brant Allen.

A friend and researcher at the University of California, Davis, Brant Allen, took this photo in Thailand as part of a photo project to document the world's largest freshwater fish. I was having trouble holding on to the fish, a common problem when handling a healthy fish in its own environment. Brant and I traveled together to many locations—Thailand, Canada, Texas, and Mongolia—to photograph giant fish that in many cases had not been systematically visually catalogued. Photo by Brant Allen.

A typical scene while filming *Monster Fish:* we would often use multiple boats, one for fishing, one for filming, and a support boat. In this scene, one of the Fishsiam guides, a man named Om, had hooked a big stingray as the sun was setting. Given that a large stingray can take hours to bring to the surface, we realized we had a long night ahead of us. When fishing and filming, we were never sure what would happen or what we would find on any given day. Photo by Zeb Hogan.

Here I am with Dean "Johno" Johnson, filming stingrays in Thailand. Johno is one of the owners of InFocus Asia, a production company that helped pitch the very first episode of *Monster Fish* in 2007. Johno produced and served as cameraman for the first show that was greenlit by National Geographic. Photo by Zeb Hogan.

Although stingrays are often considered purely marine species, there are dozens of freshwater ray species. This giant freshwater stingray was caught in 2022 in Cambodia. Photo courtesy of Chea Seila.

species on Earth, and that we have identified only about a fifth of them. But the 8.7 million figure is probably at the low end of the spectrum, with most estimates ranging between 10 and 15 million. And those estimates often overlook microorganisms, such as bacteria, of which there may actually be one trillion species on Earth.

Defining, identifying, and distinguishing between species is a complex and difficult process, and Charles Darwin didn't make it any easier when he showed that species change over time. Biologists even argue over the definition of the word "species," a disagreement that's been referred to as the "species problem."

For the most part, it's agreed that species are made up of organisms evolving together, and that shared reproduction within species and the evolution of reproductive barriers between them are major factors that cause species to exist. But these ideas only help us answer the question of what species are, not how to identify them, as I was made keenly aware during my many visits to the Thai fish markets, trying to differentiate the various catfishes by their barely visible teeth.

Interestingly, fishes were the first group of animals for which patterns of paleobiodiversity, or fossil diversity, were systematically documented, by Louis Agassiz in volumes published 1833–43. In contrast, only a small number of existing fishes had been scientifically described at that time. In part, that was because there were not a lot of people documenting life under water.

One person who did was Pieter Bleeker, a Dutch medical doctor with a keen interest in ichthyology, who in 1842 was sent to Batavia (today Jakarta) in the Dutch East Indies (now Indonesia) as a medical officer in the East Indian Army. There he founded a scientific journal focused on natural history, taking a particular interest in identifying previously undescribed fishes.

When Bleeker returned to the Netherlands, in 1860, he began publishing a multivolume encyclopedia of his discoveries. The work, which was still not completed by the time of his death in 1878, would become the most extensive by any Indo-Pacific ichthyologist, with Bleeker describing 1,925 new fish species, of which more than 700 remain valid today.

One of those species was the giant freshwater stingray, which Bleeker had described in his journal in 1852. His account was based on a juvenile specimen that he reportedly collected in Batavia. Bleeker had named the new stingray species *Trygon polylepis,* from the Greek *poly* ("many") and *lepis* ("scales"), classifying it as part of a genus of stingrays that would later come to be called *Himantura.*

In subsequent years, this stingray slithered into scientific obscurity, almost as if it had never existed, with seemingly no mention of it in the literature for more than a century. Then, in 1990, a study appeared in the Japanese *Journal of Ichthyology,* in which two scientists, Supap Monkolprasit and Tyson Roberts, reported the discovery of a new giant freshwater stingray species from Thailand, which they named *Himantura chaophraya,* because it had been described from three specimens caught in Thailand's Chao Phraya River.

The new species, the authors wrote, reportedly attained weights of five hundred kilos (eleven hundred pounds) or more (an assessment that appeared to be based primarily on newspaper accounts), and it belonged to a group of stingrays found mainly in fresh water and which probably occurred in the Mekong River as well. The study made no reference to the stingray that Bleeker had described, though it did mention *Himantura bleekeri,* a separate and smaller marine species known mainly from the Bay of Bengal.

But was it actually a never-before described species? There were those who weren't so sure, including Peter Last, an Australian elasmobranch expert, and Mabel Manjaji-Matsumoto, a Malaysian ichthyologist. The latter had seen stingrays that seemed identical to the Thai species in Malaysian Borneo, where her university is located.

Together with Last, she compared stingrays from Java, Indonesia; Thailand; Sabah, in Malaysian Borneo; and India, and found that *T. polylepis,* the species that Bleeker had found, and *H. chaophraya,* the "new" Thai species, were actually one and the same. Since Bleeker's name had been published earlier, the scientific name of the giant freshwater stingray became *Himantura polylepis.*

But the stingray's identity crisis, if you can call it that, was not over. Several years later, in 2016, Last and others would conduct a large, morphologically based review of eighty-nine stingray species in the Dasyatidae family for a guide to the world's rays and found that most of the Dasyatidae genera were not monophyletic groups; that is, they were not a group of taxa composed of a common ancestral population. A taxonomic revision of the entire family led to many of the species formerly assigned to *Himantura* being reassigned to other genera, including the giant freshwater stingray, which came to be known as *Urogymnus polylepis.*

It meant the ray had been called three different names since I had seen my first specimen. But to be honest, its scientific name was not that important to me. I was more curious about the stingray's ecology, conservation status, and size.

That's not to be dismissive of taxonomy. Taxonomy helps us categorize organisms so that we can more easily communicate biological information. It's a way for us to understand the diversity of life on our planet, which is becoming ever more crucial as that biological diversity is lost and species go extinct. The more information we have about a species, the better we can protect it. So I am thankful for taxonomists. I'm happy there are people who can do this work. And whatever they tell me to call a species, I'll call it that.

~~~

One day in 2007 I heard from Jean-Francois Helias, a French recreational angler whom I had met in Cambodia. He was calling about the giant freshwater stingray. Jean-Francois said he had teamed up with a new sportfishing outfit operating on the Bang Pakong and Mae Klong Rivers in central Thailand and that they had been catching huge rays. It didn't surprise me that sport fishers were landing stingrays. More perplexing was where the catches were made. I knew the stretches of the river that Jean-Francois was talking about. They ran through bustling towns in a semi-urban region. Could the elusive stingrays, which I had scoured the remotest parts of the Mekong to find, to no avail, really be hiding in such an area?

By this time, the accounts of giant rays had kept piling up. I had seen an old report in a Thai newspaper, the *Daily News,* of a ray weighing 295 kilos (650 pounds), two kilos heavier than the record catfish. A photo of the ray had run on the front page of the paper, and it looked huge. But the story said nothing about how the fish had been weighed. I came across another account, this one of a stingray that had been landed on the Nan River in central Thailand. According to the locals, the catch had been placed live in a dump truck and weighed on a rice scale at a local mill. The weight was reported as an astounding 485 kilos (1,069 pounds), almost 200 kilos heavier than the Mekong giant catfish. But the source of the story was a dubious one—the Thai version of "Ripley's Believe It or Not!" I chose not to believe it.

Jean-Francois told me that the anglers he had fished with routinely made catches of stingrays weighing over five hundred pounds. They had not been able to get verified weights, however, and I wanted to go there myself to see if the reports were true. This was in the spring of 2008, and Stefan, who had continued writing and filming the online series with National Geographic News, came with me.

The fishing outfit was called Fishsiam and was run by a cheerful, rough-around-the-edges Englishman named Rick Humphreys and his business partner, Wuttichai Khuensuwan, a Thai angler whom everyone called "Boy"—or "Bad Boy," as he liked to call himself. I had spoken to Rick, who offered to take us to Bang Pakong River, east of Bangkok, where his team had been fishing for the rays. We met in Bangkok one morning, where the stifling heat and heavy traffic had already set in as we inched our way out of the city.

The Bang Pakong is said to be one of Thailand's five holy rivers, and there are several famous temples situated along its course. Known as the "river full of prawns," it originates at the confluence of two other rivers, and runs for 140 miles through the densely populated south-central part of Thailand before emptying into the gulf not far from Bangkok.

Our destination was Chachoengsao, the capital of a province of the same name and through which the Bang Pakong flows. There, we booked into a hotel in the middle of the city. After dinner, I went to my room as the others stayed in the restaurant to watch a soccer game featuring Rick's favorite team, Arsenal. I could hear Rick shout at the TV, whether in approval or disgust I wasn't sure.

In the morning we headed to the riverbank on the outskirts of town, traveling in tuk-tuks. The place Fishsiam was fishing from turned out to be an old boat launch, and we were met there by Boy and a dozen other anglers and helpers. Like Rick, they were decked out in green Fishsiam shirts to promote the company, which was attracting a growing number of clients. Stefan was planning to do a video story in addition to a text installment for *National Geographic.*

The river ran murky and wide, maybe a thousand feet from bank to bank, and had large clumps of water hyacinths drifting along the surface. There seemed to be no one around to monitor any fishing activities. I had been told that local Thai people were wary of the stingrays. They called them Rahu, the god of darkness, and catching one was said to bring bad luck. No scientific research had been done on the ray population in the Bang Pakong. All I knew was that the Fishsiam crew had been catching the rays regularly, and some of the catches had been very large. But no big rays had been weighed, since that required a big scale and a lot of maneuvering to get the giant fish onto it, probably involving a crane.

We had not brought any such equipment with us. I had simply come to investigate if the reports I had heard from Jean-Francois were true. Pulling a giant ray out of the water was going to be tough enough. When hooked, the stingray would instinctively dart downward and dig into the river bottom,

Rick told me. It meant that, in addition to the considerable weight of the fish itself, several hundred pounds of mud might also be added to the load.

With the sun beating down on us, we joined Boy in a boat and went into the middle of the river where some of the team had already been fishing. They were all wrapped in scarves and bandanas to protect themselves from the bright sun. It was incredibly hot and there was no shade to escape to. With no fish biting, we decided after a while to go back to the shore.

We hadn't been back long before the rest of the group returned, bringing with them a freshly caught small stingray. The ray was no more than two feet across, excluding the tail. It was clearly a juvenile, but we still had to be careful with it because stingrays develop their venomous spines at an early age. We wrapped cloth around the stinger to make sure none of us would get stung. Although stingrays are not aggressive animals, and this particular ray did not seem agitated at all, we could not take any chances.

When stingrays do strike, it occurs in the blink of an eye, and even small wounds can be incredibly painful. And serious accidents, though uncommon, do occur. Two years earlier, Steve Irwin, aka "the Crocodile Hunter," had died after being pierced in the chest by a marine stingray while filming a documentary on Australia's Great Barrier Reef.

After taking measurements and examining the ray, we returned it to the river. Later that day, the team caught at least two more stingrays. They too were fairly small, but the catches proved there were plenty of stingrays in the river, and, unless Rick had been lying about the catches they'd made, we knew some of them could grow very large. The question was: how large?

The following day we stayed at the hotel to do some work, while Rick returned to the same location to continue fishing with his team. A couple of hours had passed when my phone rang. It was Rick. The guys had caught one, he said. They hadn't brought it out of the water yet, but it was big. We should get down there immediately.

So we jumped into a tuk-tuk. While zigzagging through the city streets, Stefan filmed me as I excitedly explained to the camera that we had got a call from Rick about a big ray. It's these moments that are often the most exciting: when I've heard of a catch of a big fish—or have one on the end of the line—but haven't seen it yet. I love the anticipation and adrenaline that comes with not knowing exactly what I'll see, but with the possibility of it being the biggest I've ever encountered.

As soon as we pulled into the boat launch, Rick ran toward us. It was a big ray, he shouted, but that wasn't the most extraordinary thing. The ray had turned out to be a pregnant female, and as soon as the anglers had brought her to the shore, she had given birth to a live baby ray. I hurried down to the river's edge, where ten guys were standing in the water, surrounding the ray, making sure it stayed in place. It was a giant, but not as big as the one I had seen in Cambodia all those years earlier. One of the men held up its tail. A towel had been wrapped around the area where its stinger was located.

The anglers had placed the tiny baby ray on top of its mother's broad form. I waded into the water as Stefan kept filming. The baby stingray was tiny, less than a foot wide but fully formed. As one of the guys pushed water on top of it, the baby flapped its edges, as if trying to take off in flight. The mother, on the other hand, did not move much. She didn't show any injuries, but she must have been completely exhausted from both getting caught and giving birth.

We knew we should not keep her too long before releasing her back into the river, especially not with the newborn at her side. So we went to work, taking measurements (190 cm without the tail) and photos. I talked to the camera about the mystery of the giant freshwater stingray, its remarkable physiology, and the extraordinary event that we had just witnessed. Then it was time to let the rays go. As the sun sank below the palm trees across the river, we pushed the baby and mother back into the murkiness of the Bang Pakong, where they disappeared in an instant. Back at the hotel, I sent a message to National Geographic describing what had happened, and later the story ended up on *Wild Chronicles,* a show to which Stefan regularly contributed stories.

Early the next morning I was awakened by a call from Cambodia. It was Sok Long. He was speaking in Khmer, but I was able to understand a few words, including "trey pabel" (stingray) and "thom" (big). We had planned to stay for one more day at the Bang Pakong, but this was an opportunity to see another giant stingray. If I could get there within twenty-four hours, he could keep it for me rather than sell it to market. I hired a car to take me back to Bangkok, and from there I flew to Phnom Penh, where I hired another car to drive me the six hours to Kratie.

Arriving at the Mekong's edge, I found a rope tied to a pole and leading into the murky water. As Sok Long pulled on the rope, it took a few moments before the dark-colored mass of the stingray's body breached the surface. We moved it onto a blue tarp, taking care to constantly put water on top of the fish. I knew I could not keep it too long in that place, so I sprang into action. The stingray measured 4.2 meters (almost 14 feet), which was slightly larger than the ray I had seen at Prey Veng.

I handed my small video camera to Sok Long, who filmed me talking about the ray, and after that he took a series of photographs of me and the ray as I was holding up its tail. One of those photos would be my favorite for many years to come; the angle at which Sok Long took the photo made the ray look even more massive than its already enormous size. That day I also happened to be wearing a bright green shirt that a friend had given me as a joke. If you look closely, you see a drawing of Nessie and below it the text "Loch Ness Lives!"

A few minutes later when we released the stingray, it quickly swam away. It was clearly a resilient animal, and I was no longer in any doubt that these fish could grow to enormous proportions. Having encountered two giants in two different river systems in two days, it seemed just a matter of time before I would find a record breaker.

The giant stingrays were made for TV. With their alien look and mysterious aura, they were a natural subject for *Monster Fish*. Now it had also been established that they could be caught, and thus filmed. My visit with Rick to the Bang Pakong River became a scouting mission for the show.

The episode about the rays was going to be produced by InFocus Asia, which employed an excellent fixer and researcher named Jatuporn Athasopa, who went by "Tuktaa." Tuktaa found several stories of stingrays being caught elsewhere in Thailand, including by one fisherman in Nakhon Sawan in central Thailand who had reportedly caught four stingrays in the Ping River in front of his house. One of the rays, a male, was said to have weighed 320 kilos, over 700 pounds, but there was no information on how that weight had been arrived at. The other three were females, and pregnant. All of the rays reportedly had been supplied, alive, to the local aquarium, where they were being housed in an outside pond.

We began production on the show in January of 2009, and at Rick's suggestion began fishing on the Mae Klong River. There we set up base at an old hotel resort situated upstream from Samut Songkhram, the capital of Thailand's smallest province, located at the mouth of the Mae Klong as it empties into the Bay of Bangkok. The river looked much like the Bang Pakong, wide and brown, but it had been diverted into many canals (*khlong*, in Thai) used for irrigation throughout the province, which is the leading salt-producing area in Thailand. Not far from our resort, several canals also ran through Amphawa, a floating market town and major tourist destination.

Boy and his team quickly made some catches. Their first few rays were small, one measuring 130 centimeters in width and 50 kilos (110 pounds). But later they caught a bigger fish—150 centimeters wide and 108 kilos (238 pounds). We were able to weigh it using a scale with a sling for the ray. It worked fine for a stingray of that size, but would not be sturdy enough for a record breaker. For filming purposes, it was a productive day, with lots of action and progressively larger catches. It felt like we were on track to find the big one. It was also helpful to get measurements and weights of rays of different sizes in order to come up with a formula for estimating weights based on the dimensions of the fish.

At the Mae Klong, we also partnered with Nantarika Chansue, a veterinarian from Bangkok's Chulalongkorn University in Bangkok. Dr. Ning, as she was known, had taken an interest in the stingrays. She and her team wanted to implant tags in the rays to track their movements, and also collect toxicology samples from the venomous mucus found on the barb of the stingrays. The samples would be analyzed with the goal of creating a vaccine to combat the effects of the toxic venom. Such a vaccine seemed needed, since there were quite a few reports of people, including some Fishsiam members, being injured by stingrays.

After several days of filming, one of the anglers hooked a very large ray, which must have immediately burrowed itself into the river bottom, because it seemed impossible to move. Strapped into a harness, the guy kept pulling and pulling, without much success, before handing over the rod to Boy, who, strapped in a harness himself, continued pulling until he, too, needed a break and handed the rod back to the first guy. On like that it went for three hours, until the fish had grown as tired as the anglers and finally relented.

As it was brought up to just below the surface, it became clear we were dealing with a massive ray. It seemed to be well over two meters in width, a potential record breaker. But then, just as we were getting ready to net the monster, the 9/0-size stainless steel hook that had been used to catch the fish snapped cleanly off at the shank, and the giant stingray disappeared into the deep.

From the Mae Klong we moved on to the Bang Pakong River, making a filming stop at the wholesale fish market in the town of Samut Prakan on the way. Housed in an open-air warehouse, the huge market sold several species of large sharks as well as marine stingrays, including cownose rays and eagle rays brought in by trawlers and long liners operating offshore. Dr. Ning and I suspected that, in the Mae Klong River, the freshwater rays were at least moving into the brackish waters of the estuary, but I didn't find any freshwater rays for sale.

At the Ban Pakong, we had arranged for a large scale with a crane in the hopes of getting to weigh a record-breaking ray. But a few days of fishing resulted in the capture of only a few small stingrays, though several of the baits we retrieved had been crushed by rays much larger in size. Then, on the final day of our stay there, the team hooked into a large ray, which it battled for over an hour and a half. It was getting dark when the ray was finally brought up, which made it difficult to film. The bigger problem, however, was where it had been caught, on the far bank of the river, a fair distance from the landing spot with the scale and too far to bring the fish to. Instead, the team had to bring it to the muddy riverbank, where we measured it to be 185 centimeters in width. According to the formula I had worked out, that meant it probably weighed around 185 to 190 kilos (or a bit over 400 pounds). It was amazing and impressive by almost any standard, but still more than 200 pounds shy of the record Mekong catfish.

We decided to go back to the Mae Klong River. At the time, the Fishsiam team was hosting an angler named Ian Welch, a doctor from the UK. People had been fishing in different groups for three days when Ian suddenly hooked into something big. An hour's struggle with what was clearly a very large stingray ensued, before the fish was finally netted and brought to the riverside. It was a female and enormous—a whopping 210 centimeters across. With the thickest body anyone had seen in a giant freshwater stingray, it seemed like we had finally found our record breaker. But there was a problem: at the Mae Klong we no longer had the large scale and crane with us.

The producers scrambled to find a way to weigh the monster ray. They wanted to bring a big scale and crane to the river. But it was going to be difficult, and even if they succeeded it was probably going to take some time. Meanwhile, as Dr. Ning and her team began examining the ray and preparing to implant a tag in the fish, another complication arose. An ultrasound examination on the ray found it to be pregnant. It meant we didn't have much time, because keeping a pregnant stingray out of the water and subjecting it to more stress was not an acceptable option, as it risked harming both the mother and her unborn young.

We decided to move the ray from the big tarp where we had placed her and into a small inflatable swimming pool. But as time ticked on and no scale arrived, it became clear we could no longer keep her there either. The reality of the situation sank in. We were not going to be able to weigh the ray. We couldn't risk killing her or her offspring just to get a weight. Yet it was frustrating to think that we could have an answer to the fundamental question driving my search. We could be looking at the world's largest freshwater fish.

I felt conflicted. I was excited by the catch and appreciative of the fact that I had a chance to see, with my own eyes, such an extraordinary fish. But I was also disappointed not to be able to get her weight, which might have confirmed the fish as the biggest there ever was. And so, with the sun sinking below the horizon, we released the giant back into the river.

~~~

The highest diversity of sharks and rays is found in Australia. In a comprehensive identification guide published in 1994, authors Peter Last and John Stevens covered almost three hundred species of chondrichthyans occurring in the seas, estuaries, and fresh waters of the region, though a third of those fishes could not be scientifically identified; that is, their existence was known but not their names. That, scientists agreed, not only presented a nomenclatural dilemma but also complicated conservation efforts, especially since the IUCN had listed some of the species, like the Port Davey skate and the speartooth shark, as threatened.

It would take more than a decade before even a quarter of the unnamed species were formally described. Among those species coming into existence, by way of a 2008 paper by Last and Manjaji-Matsumoto, were seven stingray species, six of which were found in marine habitats.

The sole nonmarine species was a freshwater whipray occurring in rivers in northern Australia. This ray had not been known to science until a specimen was collected by harpoon in 1989 during a scientific expedition on the Daly River. After initially being mistaken as a regional subpopulation of the similar-looking giant freshwater stingray, the ray was given the scientific name of *Himantura dalyensis,* after the river where it was discovered.

I heard about this new freshwater whipray species, and I got in touch with Dion Wedd, an aquatic ecologist who was working at the Territory Wildlife Park, a zoo-style nature reserve located south of Darwin. The park, I was told, was home to several whiprays, and they were large specimens, though not as big as the giant stingrays I had encountered in Thailand and Cambodia. Deon, however, believed the whiprays grew to similar proportions, at least up to two meters wide. I was naturally keen on investigating, and Nat Geo was also excited to feature the whiprays for a *Monster Fish* show.

It was the middle of the dry season when we flew to Darwin, the largest city in northern Australia. Although it is the largest city in the region, Darwin has a Wild West feel to it, and you don't need to stray too far outside the city limits to be in the outback. When we arrived, it was hot and dusty, much different from Darwin in the rainy season, when, I was told, it was similar to Thailand and Cambodia, with torrential downpours turning shallow creeks into muddy, rainfed rivers.

We met with Dion in Darwin, and together we began the long drive out to the Daly River, where Dion thought we'd have the best chance of finding a big whipray. He told me of a huge ray he had seen in the Daly many years before, one that rivaled the giant freshwater stingray in size. I had a suspicion he was prone to telling tall tales, but I was excited to see the river for myself.

The drive was mile after mile of red, dusty roads, shrub, eucalyptus trees, and cattle ranches, or stations as they're called in Australia. We finally reached the Daly River crossing. In the dry season it is possible to traverse the river by car on a narrow cement roadway that bisects a wide and shallow section of the waterway. The river itself was an oasis with plenty of water spilling generously over rocks, cobbles, and sand and winding through what otherwise would be hot, inhospitable land. It reminded me of the clear canyon streams back home.

The Daly River, I was told, is one of the only perennial rivers in northern Australia. It is spring-fed, and in the dry season, without heavy rains or runoff to cloud the river, its clear waters form an intricate network of small, vegetated islands, sandbars, pools, and knee-high waterfalls. In that sense, it is similar to the Fitzroy, where we had filmed the sawfish episode. But while the Fitzroy turns into a series of isolated deep pools in the dry season, the Daly keeps its flow and is full of movement, with small riffles and rapids, water flowing over ledges and over soft sands in some places only a few inches deep. It's also home to a wealth of wildlife, as I would quickly find out.

After the hot and dusty journey, I couldn't wait to get into the water for a swim. Coming from Arizona, I was used to jumping into a river to cool off after a long car ride, and as soon as we stopped I headed down to the riverbank. I was about to jump into a swimming hole just below the crossing when I looked down into the river and saw movement. A half-dozen dark shapes circled in the waist-deep waters. At first, it didn't compute in my head what they were. Here we were, in the middle of the desert, a hundred kilometers from the ocean, and there was a school of bull sharks swimming below me. I did not jump in.

We set up our base camp at the crossing with a plan to spend our days exploring the river just downstream. This stretch of river seemed like good hunting ground for us because the road crossing acted as a barrier to fish movement during the dry season, and so many of the larger aquatic animals congregated in the area. But as we found out, this also presented a problem when trying to catch the whiprays. Every time we put a line out, we caught a bull shark instead, usually a small one, just a few feet long, but occasionally something larger that would take our bait and cut our line.

There were plenty of other large fish too, like barramundi. Whiprays, on the other hand, seemed far less common. It was not until we headed out at night with powerful flashlights that we finally stumbled on one. It was small, about a foot in diameter, and seemed to be exploring the shallows, probably looking for food. We scooped it up with a drift net and examined and filmed it before releasing it back into its environment.

Later on, and farther downriver, I was fishing when I hooked into what was clearly a much larger ray. Although I had a heavy-duty fishing pole and was using a 100-pound braided line, I couldn't budge the fish. Given the current in the river, I didn't want it to make a run into deeper water, and that's when I made an amateur mistake by fully tightening the drag on my fishing reel. I thought I could keep the fish close to the riverbank by sheer strength and force of will, but of course the ray had other ideas. A soon as I tightened the drag, it made a run to the middle of the river and the braided line broke.

A 100-pound braided line is strong—a full grown man could probably hang from it—but not strong enough to corral a ray that uses its mass and the water current to quickly pivot away. On some level I knew it was silly to be using a fishing pole and braided line in a shallow, multichannel river to catch a fish as big a whipray, but it's hard to imagine the strength of the fish until you actually hook into it. It was why Sok Long used huge hooks, thick rope, giant floats, and a massive rock as a weight to catch the giant freshwater stingray in Cambodia. A rod and braided line was not going to cut it. The gear had to match the fish.

While the presence of the bull sharks complicated our efforts, the sharks were no real danger to us. It was different with another inhabitant of the river and a much larger predator: the saltwater crocodile. We knew salties were in the river because we had seen one sunning itself on the opposite bank. We could also see their eyes at night, reflecting from our flashlights as we scanned the river looking for rays. When it came to swimming in the river, I had to abide by the rule that if I could see the bottom of the river, it was okay to get in. In spots where I couldn't see down deep, I would stay on land.

As we learned, the best place to swim in the river was in the shallows where we had caught the small ray early on during the trip, and where it became clear that other juvenile rays would gather. The water in the shallows was perfectly clear and just deep enough to swim in. Best of all, it was free of crocodiles. Donning a mask and snorkel, I could submerge and swim next to the small whiprays as they explored the sandy river bottom for food.

Swimming with the rays became the highlight of the trip. The young rays had a lot of personality. While most species of fish, even large ones, will quickly retreat from an encounter with humans, these baby rays swam toward me and lifted their snouts as they got near. I assume they were trying to get a whiff of me and to figure out exactly what I was or if I had food. Once they lost interest in me, they swam off under cutbanks or buried themselves in the sand. Watching them shake and sink into the sandy bottom was cool; one minute they were in front of me and the next minute they were gone, or at least hidden under the sand. It partly explained why we had such trouble finding larger whiprays. Rather than swimming away from predators, the rays would bury themselves, becoming invisible. Given this behavior, it was hard to know how many whiprays were in the area.

Swimming with the rays reminded me of how rarely we get to see fish in their natural environment, rather than after they are caught and exhausted or dead. It is altogether different to see a wild fish, healthy and unhindered. Observing the way they move through the water, using the current to their advantage and seeking out food and shelter, changes our perceptions of the fish. It is similar to watching a crane in a courting dance, a pack of wolves on a hunt, or a mother chimpanzee caring for her young. Swimming with the rays was a chance to appreciate these animals on their terms, for the wild animals they really were.

Our visit to the Daly River also produced one of my favorite photos. It is of me filming a small whipray and a bull shark in one of the shallow pools. We had caught the two fish in a nearby section of the river and placed them in the pool to film them close up. I love the photo because the river looks just like Havasu Creek, a spring-fed creek in Arizona where I worked in college and which is one of my favorite places on Earth. Except in this photo, a whipray and a shark swim in the water on each side of me. The whole scene seems so incongruous. It reminds me of how special these places are and how lucky I've been to experience them.

In the end, it was impossible to verify the stories that Dion had told me about seeing Australian whiprays more than two meters wide. Were his stories true or was he just having a laugh at my expense, knowing how hard it would be for me to ignore his tales about record-breaking rays?

Most of the world's freshwater rays are found in rivers in South America, and among them is a giant ray that is a serious contender for the title of world's largest freshwater fish: the short-tailed river ray (*Potamotrygon brachyura*), which gets its name from a short and powerful tail as thick as a man's leg and is said to grow up to three hundred kilos, which would be record-breaking territory.

The river ray lives in the Paraná River, the second-longest river in South America, which runs for three thousand miles through Brazil, Paraguay, and Argentina. But details of its distribution, growth, and overall ecology are fuzzy at best. In relation to the other two ray species in this chapter, it is the least understood, even though those other two species were not properly identified by science until recently.

In my attempts to research the short-tailed river ray, I couldn't find much scientific information beyond a few lines on FishBase, including a published max weight of 208 kilos (458 pounds). On the IUCN Red List, the ray was described as "data deficient," with the latest assessment carried out in 2003. Even a decade into the project, I was amazed at how little scientific information could exist about one of the largest river creatures in the world.

As with other megafish, the most useful information came from fishing accounts. I found reports of large river rays caught around the city of Bella Vista, a summer tourist destination on the Paraná in Argentina, near the Paraguayan border. I was also contacted by a man named Mark Jones, who knew about the rays. Mark had an aquaculture background as an oyster farmer in his native New Zealand and later in Chile, where he also farmed mussels. More recently, he had started an angling and tour operator business in Bella Vista, and he himself had caught a ray weighing an IGFA-verified 218 kilos (480 pounds). He also had a credible account of a ray caught down-stream, around Santa Fe, that weighed 260 kilos (573 pounds). He believed there were rays growing even bigger than that.

Mark offered to facilitate a fishing expedition if we wanted to come and film *Monster Fish*. He included several photos of huge rays that had been caught and, to spice up his pitch, wrote of the power of the ray's tail: "I have seen horses with both legs broken by the tail of a large ray." It was a promising lead, and I was hoping that InFocus Asia could help make it happen. It took more than a year to work out the logistics, but finally, in early 2015, we were ready to film the show. I boarded a flight to Buenos Aires to meet the crew, which included producer Robbie Bridgman and cameraman Rob Taylor.

It was my first time in Argentina, and it proved to be a demanding trip from the start. From Buenos Aires, we took an overnight bus to Bella Vista. The central bus station was a madhouse, with a lot of confusion about which bus was going where. We also had something like twenty-five bags of gear to load. The bus trip ended up taking more than ten hours, in part because of the torrential rains that completely flooded the roads. At the bus stop in Bella Vista, we were met by a woman with a small car who had to make four trips to take us and all of our gear to the hotel.

I had not realized how big the Paraná River is. (I would have been better informed about its size had I spoken the Tupí language, since the name Paraná is an abbreviation of the Tupí phrase "para rehe onáva," which means "like the sea.") In sections, the river is more than ten miles wide, though midchannel islands make it difficult to fully appreciate its enormity. It has a harshness to it too, with turbulent waters, bugs, heat, and frequent fast-moving thunderstorms.

Anglers came to the Paraná mainly to fish for dorado and surubim catfish, not stingrays. In the few years since Mark had started his fishing outfit, only a handful of clients had fished for short-tailed river rays, despite their impressive appearance. There was a reason for that. As we soon found out, the huge rays were difficult to catch, and fishing for them required patience and tenacity.

The best place to fish for the river rays was downstream from Bella Vista, about an hour by boat, so that's where we headed with our guide, Olfi Pesca, a happy-go-lucky Argentinian with an infectious enthusiasm and enough experience fishing for rays to give us confidence we could find them. He had been with Mark when he caught the 480-pound ray, and had accompanied angler Jeremy Wade on his quest to find the fish a few years prior to my trip.

Olfi knew this stretch of the Paraná like the back of his hand. He knew the river rays liked to stay near seams in the river, around eddies, in deep holes, and close to sandbars. He also knew that, in contrast to the aggressive taimen, the rays' hunting behavior was reserved and cunning. The rays would not pounce on the bait (which consisted of chunks of fresh fish) like the taimen did, but instead examine and play with it before possibly biting down on it. It was essential to be patient and wait to see a bend in the rod before trying to pull up the fish.

But the river was also chock-full of yellow-bellied piranhas, which inevitably devoured the bait before any ray could get to it. Time and again, I would reel in the line only to find an empty steel hook. It was exasperating, and things were not made any better by the conditions out on the river. Swarms of mosquitoes and biting flies made it hard to concentrate on fishing or even filming, since the clouds of bugs were so thick. The weather could change in a minute, with the winds whipping up and sudden downpours making life generally miserable.

After fishing for a week in the daytime, we decided to try our luck at night. Being out at night had its drawbacks, however. It was less safe and not particularly good for filming. There were also just as many bugs, and the weather could turn just as quickly, making for treacherous conditions. As another thunderstorm moved in, big waves pounded our boat, which filled with so much water I thought we might be in danger of sinking. On several occasions we had to abandon our fishing efforts and head back home. It was one of the most physically challenging of all the *Monster Fish* shoots, and I felt for Rob, who always had it the hardest, having to keep the gear working while capturing useful footage.

But then things turned around. We were joined by a few more people who could help us fish, and after several days on the river I was starting to get the hang of how to fish for these rays. At first I caught some smaller rays, and then one that measured three feet across the disc. I learned to differentiate the bites of the piranha from those of rays: the rays typically took the bait slowly, then either sat still or made a long, steady run.

Once a ray was hooked, the difference was obvious. They were powerful animals in the flowing water and took serious work to bring to the boat. Fortunately, with the sandy riverbanks it was easy to bring the boat and the rays to shore, unhook them, and film them without too much struggle. We could keep the rays in the water the entire time without worrying about other predators, like saltwater crocs. I would get in the water too. The smaller rays were beautiful and perfectly matched to their sandy environment.

With each passing day, we put in longer and longer fishing hours trying to catch a full-size ray. Olfi enlisted some more friends to help us catch bait and to find good fishing spots. With just a couple of days left of the trip, we were fishing in a large backwater lagoon covered in water hyacinths when suddenly I hooked a very big ray. As I was using a smaller rod and less durable line, I was afraid I would lose it. But the lagoon was not deep and seemed to have a sandy bottom, so the ray could not nestle its way under some snag that would break the line.

However, in the dark, with the murky water, vegetation, and camera lights, it was hard to see what was happening. I had to feel the action of the ray through the rod and keep it from going under the boat or getting too far away from us. Eventually, the fish tired out and I was able to bring it to the surface. It was a gargantuan creature, not as wide as the giant stingrays in Thailand and Cambodia but with a much thicker body and tail. After a long struggle, we were able to get a bag net under the fish and unhook it. The entire group let out a collective sigh of relief. We knew that until we had the ray in our net it could easily escape and we wouldn't have had the opportunity to measure or film it.

By this time it was 4:00 a.m. and would be light soon. We were all exhausted and made the decision to rest a bit and wait for dawn before filming the release. We took the ray and the boat to the riverbank and set out blankets on the sandy shore to wait for the light. After an hour had passed, we were still tired but felt a little better. Wading into the water, we had our first good look at the ray. It was a beast, a healthy female, darkly colored with a beautiful spotted pattern and a tail as thick as Olfi's leg. Some of the other rays we had seen were a bit tattered around the edges as a result of piranha bites, but not this one.

Before examining it more closely, we wrapped the spine in cloth and stayed clear of the tail in case the ray became agitated. Rob got some great footage of the fish as I handled it. We had no way of weighing it, but from carrying it we estimated it probably weighed about four hundred pounds. It was the biggest river ray that anyone had ever filmed. In the end, the whole process went smoothly, and we spent only about fifteen minutes with the ray before releasing it back into the river. I had come to love the way rays swim off when released. They don't struggle or fight but rather slowly glide away as if our time with them never happened.

By this time I had become convinced the record-breaking fish would be a stingray, possibly the short-tailed river ray. This species appeared to be doing well in the Paraná River, where it was largely left alone by commercial fishers, and where it seemed to move freely throughout the river system in order to complete its life cycle. Those were the kinds of conditions that one imagined would allow the world's largest freshwater fish to flourish and grow. But there was no real evidence that the South American river rays could actually attain record-breaking weights.

It was different with the giant freshwater stingray in Southeast Asia. There was plenty to suggest that this species grew well into record-breaking territory. I had been receiving regular updates about very large rays continuing to be caught by Fishsiam in the Mae Klong River in particular. In early 2015, Dr. Ning and her team were visiting when Boy and his crew caught a stingray measuring 2.4 meters (almost 8 feet) in width and 14 feet from tip to tail. It turned out to be a recapture of a ray that Fishsiam had caught six years earlier and tagged.

Comparing the data with that from the earlier capture indicated that the ray had grown more than a foot in that time, a healthy growth rate. It was possible that this ray, too, had been a record breaker, but once again there had been no scale on hand big enough to establish if that was the case.

It was time to settle this question once and for all, and the only way to do that was to go back to the Mae Klong with gear capable of weighing a stingray of record proportions. There were also other questions to be answered about the giant stingray. Despite the frequent catches, we still had virtually no scientific data on the rays' population density, abundance, or movement patterns. This information was critical to developing effective management plans for the stingrays, which had become listed as endangered in Thailand. The primary threat was thought to be pollution, though there were also increasing reports of an illegal harvest of stingrays for the aquarium trade.

With my help, Dr. Ning and her team applied for and received a National Geographic grant to do a first-ever scientific study of the stingrays to answer questions such as: How many stingrays occurred in the lower Mae Klong River? How fast did the stingrays grow? How long did they live? What were their basic movement patterns and habitat requirements?

Dr. Ning and I were particularly interested in finding out how strictly confined the stingrays were to a freshwater environment. Giant stingrays occurred way up the Mekong River with no access to the ocean, so they were certainly capable of living their entire lives in fresh water. But there were also reports of captures near river mouths and in brackish water, so it seemed likely they could tolerate salt water to some extent, and that some populations may even benefit from movement between fresh and salt water either for reproduction or feeding.

By this time, Rick had left Thailand and moved back to the UK. He was still connected to Fishsiam, but Boy was running things on the ground now. The future of the angling business was unclear, since government regulations were put in place to prohibit the fishing of the giant stingrays amid growing concerns about their status. However, it was still possible to fish for them for scientific purposes. So one morning in March 2016 Stefan and I were joined by Dr. Ning and her team in Bangkok as we headed back to the Mae Klong for what I hoped would be a breakthrough effort at landing the world's largest freshwater fish.

This time I rented a heavy-duty floor scale capable of measuring weights up to three thousand kilos, and purchased a custom-made, industrial-strength holding platform that a metal shop in another province had built. The idea was to place the stingray on the platform, and the platform on the scale, measuring the weight as quickly as possible before getting the stingray back into the river. The advantage of this technique was that we didn't need to rent a crane or figure out how to get to the river's edge. The holding platform also held a small amount of water that reduced stress to the ray.

When we got to the hotel resort the scale had already been delivered and set up on the ground next to the boat ramp. With the rains still a couple of months away, the water level in the river was very low, and Boy was concerned that it would be difficult to move a giant ray, if we were to catch one, the short distance from the river to the scale. It might have been better, he suggested, if we had constructed some kind of floating platform that could be taken out on the water rather than have to transport the ray to the river's edge.

To improve our chances of catching rays, he had hired four more people to help with the fishing. Because of the blistering heat, they would split up into teams and take turns on the river, fishing from several boats, while we stayed on shore with Dr. Ning and her team, ready to receive the rays. It didn't take long before one of the boats returned with a small ray. The methods were the same for each ray that was caught: we measured its width and length, took blood and DNA samples, got a weight, and then tagged the fish subcutaneously with an AA battery-sized acoustic tag. The tags emitted a coded signal that would enable us to individually identify each ray using underwater receivers placed strategically along the river. The weighing platform worked well for the smaller rays. They were easy to handle, fit on the platform, and could be weighed quickly and then released.

But for several days all we caught were small rays. I was becoming increasingly worried. During previous outings there had been a day here and there when we had not caught any adults, but then the following day we landed a monster. This time, there were no big catches. Maybe the fishing of the rays had taken a toll? After all, others besides Fishsiam were fishing for rays. Maybe they were not as concerned about catch-and-release practices. Maybe the effects of the increased pollution in the river, which Dr. Ning and others had warned about, were setting in?

Or maybe we were just unlucky. Fishing for the stingrays—and most megafish for that matter—was still a highly imprecise science. It's something that any fisher understands. Some days you're simply not successful no matter what you do. Hopefully the big rays were still down there, at the bottom of the river. Yet as we headed back to Bangkok, I couldn't help but think the opportunity to confirm the status of the giant freshwater stingray as the world's largest freshwater fish had been missed. Over the years, there had been many rays caught that seemed likely record breakers, but no one had been able to weigh them. I wondered if I would ever get the chance to do so again.

Later that year, word reached me that there had been an accident along the Mae Klong River—some kind of chemical spill from an ethanol plant located not far from where we had been fishing for the rays. About seventy rays, all of them adults and many of them very large, were found dead, floating down the river like crashed UFOs.

After rushing to the Mae Klong to figure out what had happened, Dr. Ning and her team could determine only that the water in the river was slightly more acidic than normal. Later, however, it came out that considerable amounts of cyanide, contained in residue from fermented cassava used to produce the ethanol, had accidentally been released into the river. It had been enough to kill off all the largest rays, though smaller rays appeared unscathed. A court case would find five defendants guilty and ordered to pay 50,000 baht ($1,500) in fines.

After the accident, the Thai authorities clamped down on any stingray fishing activities not strictly for scientific purposes, and the Thai subpopulation of the giant freshwater stingray was listed as critically endangered on the IUCN Red List. The results from Dr. Ning's project were not easily interpreted, and the study was never published.

By this time, Dr. Ning was convinced there were no big stingrays left in the Mae Klong River, and maybe she was right. Maybe the creature the Thai people called Pla Gra Ben Rahu, which got its name from one of the terror-inspiring gods, had itself succumbed to the uncleanliness and falsehoods of the humans it was meant to torment.

Blue Heart of Europe

In the fall of 1956, US President Dwight D. Eisenhower used the occasion of a White House conference to propose a new citizen diplomacy initiative known as Sister Cities International. The idea was for cities throughout the United States to partner with cities around the world to create special bonds between Americans and people of different cultures in an effort to promote peace and prosperity.

My hometown of Tempe came to partner with the German city of Regensburg. The two cities started a student exchange program, which sent American students from Tempe's four public high schools to Regensburg each year, with German boys and girls making the opposite journey. The program was well regarded and free to participants, so when I reached my junior year of high school I applied. My application was approved, and in the summer of 1990, the year after the fall of the Berlin Wall, I left Arizona to spend six weeks in the fourth-largest city in the southern German state of Bavaria.

I had not been to Germany before, and didn't know much about Regensburg beyond the fact that it was a medieval city of similar size to Tempe (150,000 people) and located in a region famous for castles, beer, traditional clothes like lederhosen, and luxury cars like BMWs. In fact, Regensburg is considered to be the only authentically preserved large medieval city in Germany. From its founding as an imperial Roman citadel, it's been the political, economic, and cultural center of the surrounding region. It's also the oldest city along the Danube River, Europe's second-longest river, which runs through the town.

My exchange student "brother" Mathias and his family were gracious hosts and made me feel very welcome at their house, which was located outside of town in a rural area with rolling hills covered in wheat. When I first arrived, it was early summer, and Mathias was still in school. Left to my own devices, I would take the bus into Regensburg, where I explored the cobblestoned old town and its famous St. Peter's Cathedral before ending up at what came to be my favorite lunch spot: a collection of picnic tables on the banks of the Danube where you could order and eat delicious grilled sausages and watch the river flow by.

It was a fairytale setting, and I felt like I was transported back to the Middle Ages. What I didn't know was that in the Middle Ages, or at least before the advances of industrialization, the Danube River actually looked nothing like it does today. Similar to most of the rivers in central Europe, it was, for much of its course, not a single-channel waterway, like the one I was admiring from my seat at the picnic table, but a sprawling network of channels connected to vast and biologically diverse wetlands. And in that river network once swam huge fish, like Europe's largest trout, the huchen (or Danube salmon), and no fewer than six species of sturgeon, including the world's biggest, the beluga, which had long since disappeared from the upper reaches of the river.

Before we get into what happened to those fish, let's talk about the river itself. Because the Danube teaches us a lot about the history of European rivers and how many of them lost their natural functions to the point where Europe is now the continent with the world's most environmentally degraded waterways.

The Danube is a river of immense importance. In fact, the European Commission recognizes it as the single most important non-oceanic body of water on the continent. As the only major river in Europe that flows from west to east, it originates in Germany's Black Forest and ends in the Black Sea. Picking up two-thirds of its water from the northern Alps, it runs through ten countries (with its basin extending into another nine), making it the most transnational river in the world. With four European capitals situated along its shores, the Danube serves as a vital commercial highway for freight transport, while supplying water to industries, agriculture, and the eighty million people living in the basin.

For millennia, conflict raged along its banks. It was on the Danube that Roman soldiers staged a botched mutiny following the emperor Augustus's death in AD 14, and where Roman troops would engage in battle for centuries to come. It was on its shores, near Ulm, that Napoleon crushed a formidable Austrian army in 1805, only to face defeat by the same Austrian foes four years later. And adjacent to the river, at Mauthausen, the Nazis built one of their most brutal concentration camps, which claimed the lives of more than one hundred thousand people. Conflict, albeit of a far less violent kind, has even besieged the very origin of the Danube, with two municipalities in Germany's Black Forest locked in a decades-long rivalry over the claim to call itself the official source of the river.

Yet it has not been strife, armed or otherwise, that has stripped the Danube of the wonderful wildness that once prompted Johann Strauss to write his famous waltz "An der schönen, blauen Donau" (commonly referred to as "The Blue Danube" in English). Instead, it was the advent of industrialization in the late 1800s—and the resulting construction of dams and diversions, deforestation, pollution, and the erosion of land—that drained away the river's natural features, disconnecting it from the flood-plains where black poplar trees and beavers once flourished and upending the natural living conditions for all the creatures in the river, including fish.

Despite that, the Danube remains the most fish-rich river in Europe, with more than one hundred species found there. Actually, more than half of all the fish species in Europe occur in the Danube basin. The river is home to pike, ruffe, zander, burbot (the only gadiform, or codlike, fresh-water fish), tench (a carp known as the doctor fish), and the common nase. It's also the native habitat of the huchen, a freshwater salmonid that can grow extremely large, up to 130 pounds, although such sizes have not been observed in a long time in the species, which is now found only in a fraction of its historic range and considered endangered.

Most striking, however, is the Danube's exceptional diversity of stur-geons, the group of ancient fish that has the unenviable distinction of making up the most endangered animals in the world. Historically, six stur-geon species migrate from the saltwater habitat of the Black Sea into the Danube, including the enormous beluga, but it should come as no surprise that none has been seen in the middle or upper reaches of the Danube in a very long time.

In fact, all but one of these six European sturgeon species have become critically endangered. At least one, the Atlantic sturgeon (also known as the European sea sturgeon), which FishBase suggests can grow over fourteen feet long, is believed to have already gone extinct, and researchers fear that another species, the ship (or bastard) sturgeon, may soon follow suit. The one sturgeon not listed as critically endangered is the sterlet, a relatively small sturgeon that is nonetheless considered vulnerable and is dependent on stocking programs for its survival.

In contrast to most of the large and threatened fish around the world, the decline of the sturgeons in the Danube began a very long time ago. There is plenty of documentation showing that sturgeons were being overharvested as far back as the Middle Ages. Chronicles from the eleventh century describe how beluga sturgeon provided important rations for Austrian troops marching along the Upper Danube. So-called sturgeon fences, placed across the river channels, were a common catch method hundreds of years ago and had the potential to eliminate whole spawning runs of migrating fish. Records from 1548 show fifty metric tons of fresh sturgeon meat being sold in just a few days at the fish market in Vienna.

In time, the European sturgeons took on the status of regalia and came to be offered to aristocracy and the church, while their numbers continued to decline. Eventually, large catches became rare in the upper and middle parts of the Danube, while fish from the lower delta were being harvested for the burgeoning caviar trade. And then the wheels of industrialization were set in motion.

With the river dredged and straightened to improve shipping, diversions abstracted more and more of its water. The construction of dams ramped up in Germany and Austria, where nearly sixty dams were built along the first six hundred miles of the Danube, generating hydropower that played a vital part in the industrial recovery of both countries after World War II but blocked the migratory paths of the sturgeon. The death knell came in the 1970s with two giant hydroelectric dams known as the Iron Gates, built in a gorge on the boundary between Romania and then-Yugoslavia. The dams, which remain the largest ever constructed in Europe, made any fish passage up the Danube impossible.

When I think of the Danube today, it's not the wholesome image of the river that I remember from my time in Regensburg. I know the Danube is now a sickly shadow of the river it once was. Eighty percent of its 1,770-mile course is regulated, and over seven hundred dams and weirs have been built along its main tributaries. I see a river once vibrant with life and fish that, like other rivers in Europe, has suffered a shocking decline in migratory fish populations. I see a river that's been so significantly altered for so long that people have forgotten what it once looked like and the magnificent fish that once called it home. I see a river that is unlikely to be restored—indeed, cannot be restored—to anything resembling its natural state. But maybe I'm not seeing the whole picture?

~~~

Natural regeneration is the process by which juvenile plants or trees that have established naturally replace those that have died or been killed. Forest conservationists speak of two types of natural regeneration: the spontaneous kind, where vegetation regrows from self-sown seeds, coppice, or root suckers without human intervention, and that which is assisted, where people may help trees naturally recover in various ways by eliminating barriers and threats to their growth (without actually planting new trees).

In either case, the process allows for a forest to regrow and even return to its natural state. But natural regeneration does not work if, instead of leaving the land to do its thing, we put a bunch of cattle on it or convert it to farmland. In that case, the forest that once flourished on the land will have no chance of restoring itself. It appears to be permanently lost, much in the way that rivers and their ecosystems appear to be permanently altered once we build dams on them.

When I grew up, I viewed the dams on the Colorado River as permanent, almost natural features of the landscape. The way the river looked and behaved with the dams on it was, in a way, more real to me than what it would have been without them, because I could not envision the river in its natural state. I had never seen the Colorado River fully wild. It was not until I got older and began visiting rivers that were undammed that I truly began to understand the transformations that rivers undergo when they are regulated, and what is lost when that happens.

I would imagine that people who were there when the dams on the Colorado and other American rivers were built also viewed them as permanent features of the landscape. The dams were seen as engineering marvels, indestructible, and meant to last in perpetuity. We know that little thought was given to the impact of dams on fish and the ecological functions of the river; the public was instead transfixed on the economic and technological progress that dams represented.

By 1940, hydropower provided 40 percent of America's electricity supply, and over the next couple of decades dams continued to be built at a rapid pace in the United States. But things changed as oil and natural gas production became cheaper, surpassing coal as the leading energy source, while new forms of energy also emerged, including nuclear power. By the 1960s the hydropower boom was ebbing, in part, it should be added, because many of the rivers suitable for exploitation had been dammed by then.

It had also become obvious that hydropower plants were, in fact, not indestructible, but would eventually run into problems requiring costly rehabilitation. Reservoirs would silt up, resulting in diminished energy outputs, and infrastructure needed to be repaired to avoid potential catastrophes. It was clear that the environmental damage that dams inevitably caused was often far greater than previously acknowledged. The cost of upgrading dam safety systems or keeping outdated hydroelectric equipment running decades after a dam had been installed was, in some cases, not worth it. And so, by the 1980s an idea had begun to spread: maybe some dams should be removed.

One dam that was ripe for removal was the Edwards Dam on Maine's Kennebec River, situated forty miles upstream from the Atlantic Ocean, in Augusta, the state capital. Built in 1837 to ease navigation and harness energy for the timber industry, the Edwards Dam had decimated the Kennebec's populations of migratory fish, including Atlantic salmon, several species of herring, alewife, and shortnose and Atlantic sturgeon. Its economic benefits had long since dwindled, with the dam generating a modest 3.5 megawatts of electricity annually, less than one-tenth of one percent of Maine's power supply.

With the dam's operational license set to expire in 1993, several environmental groups banded together to advocate for the removal of the dam. Two previous developments provided them with ample ammunition for the fight. Not only had the state of Maine also called for the dam's removal as part of a plan to restore the river, but in 1986 the US Congress had amended a law governing the licensing of hydropower dams by requiring "equal consideration of power and non-power values," such as fish, wildlife, and recreation.

In the battle that ensued, the coalition produced seven thousand pages of documentation on the negative impacts of the dam and the economic importance of restoring the river's fishery. They gained the support of Maine's governor, Angus King, who said the removal of the dam would help the Kennebec "reclaim its position as both an economic asset and an ecological miracle."

The proponents of the dam countered that a removal would be too expensive and would flood and destroy properties located downstream. But that turned out to be a losing argument, because in 1997 the Federal Energy Regulatory Commission voted to deny the license renewal for the dam. It was the first time it had used its authority to deny a permit against the wishes of a dam owner. On July 1, 1999, more than a thousand people in Augusta lined the banks of the Kennebec to witness the removal of the 25-foot-tall and 917-foot-wide Edwards Dam.

Environmental advocates promised that the devastated fisheries would return after the dam was gone and that the city of Augusta would benefit from new recreational opportunities and the revitalization of the city's riverfront. But even they were surprised by the speed of what happened next. As soon as the dam came down, the river rebounded and fish began to return. For the first time in over one hundred years, Atlantic sturgeon swam past the former dam site. The alewives returned, and with them everything that eats them, from river otters and osprey to bears and bald eagles.

The successful removal of the Edwards Dam had a ripple effect throughout the nation, boosting efforts already under way to take out larger hydroelectric dams in other parts of the country. One such effort was in Washington State, where environmentalists had been lobbying for two large dams on the Elwha River to be removed.

Originating in the mountains of Olympic National Park, the Elwha River flows for forty-five miles, much of it through lush forests, until it meets with the Strait of Juan de Fuca. The river provides extraordinary salmon habitat and has historically hosted runs of more than four hundred thousand adult salmon each year. But these runs, which were culturally and economically significant to many Native American groups—and specifically the Lower Elwha Klallam Tribe, who relied on the fish for their sustenance—came to an end in 1913 when the 108-foot-tall and 450-foot-wide Elwha Dam was built near the mouth of the river. It was followed by the even taller Glines Canyon Dam, constructed farther upstream.

The dams, built to generate electricity for industries on the Olympic Peninsula, including lumber and paper mills in the town of Port Angeles, blocked the migratory route of the salmon and trout that the river supported. After the dam's construction, fewer than three thousand salmon would return each year to spawn in just five miles of habitat from the mouth of the river to the dam's base.

Fast forward to the mid-1980s, when the Elwha Klallam Tribe, having lost their lifeblood of food and tradition, joined with several environmental groups to push for the Elwha dams to be removed. At that point, the dams had outlived much of their economic usefulness, with the Olympic Peninsula long since connected to the regional power grid. Still, it came as a bit of a surprise when the US Congress in 1992 authorized the federal purchase of the two dams from the timber companies that owned them and commissioned a study about removing them.

A prolonged battle followed, with the timber industry and some of the local communities opposed to the plan. It was not until 2004 that the tribe, the National Park Service, and the city of Port Angeles finally reached an agreement on the removals, with the US Department of the Interior purchasing the dams for $29.5 million. Even then, the engineering challenges of carrying out the largest dam removal in history remained. For example, what would happen to the twenty-seven million cubic yards of sediment trapped behind the dams?

The plan that was settled on called for the dams to be taken down in several stages, allowing for a gradual release of sediment. Finally, on September 15, 2011, a barge-mounted excavator began chipping concrete off the face of the 210-foot-tall Glines Canyon Dam, with the demolition of the Elwha Dam beginning later that week. The removal method used a process in which the dams were "notched down" on alternating sides, creating temporary spillways to drain the reservoirs. It would take three years before both dams had been completely taken down. By that time, salmon were already migrating past the former dam sites and into the Upper Elwha watershed, where they had not been seen for a hundred years.

But the dam removals did not just restore the river channel itself. The sediment that had been trapped by the dams, and which was now allowed to wash downstream, helped rebuild the area around the Elwha's outlet to the sea. More than seventy acres of new beach and riverside estuary habitat were created in this so-called nearshore environment, crucial habitat for fish transitioning between the river and the sea, and important feeding grounds for salmon, killer whales, and many kinds of birds.

It was a story that repeated itself with other dam removals: once the dams had been taken out, the rivers bounced back more quickly than anyone expected. The ecological rewards were nearly instantaneous. Assisted by the same human hand that had altered it, nature was able to rapidly and naturally regenerate what was once believed to have been permanently lost.

In 2019, I was one of many coauthors on a global study published in *Nature* and led by Günther Grill, a geographer at McGill University in Montreal, and my old colleague and friend Michele Thieme, who was now the lead freshwater scientist at the World Wildlife Fund in the United States. Using satellite imagery and other data, a team of thirty-four international researchers mapped 7.5 million miles of rivers worldwide to assess the rivers' connectivity, the commonly adopted term to describe the degree to which matter and organisms can move freely in a river system.

The study, which was the result of more than a decade of data processing and preparation, showed that only about one-third of the world's longest rivers remain free-flowing, meaning they have not been dammed or disrupted by humans. While there are a few exceptions, like Asia's seventeen-hundred-mile-long Salween River, most of the remaining free-flowing rivers longer than six hundred miles lie in remote regions of the Arctic and in the Amazon and Congo basins.

The situation is particularly dire in Europe, where there are no major free-flowing rivers left outside of Russia, and where migratory fish populations have plummeted. A 2020 report drawing on the Living Planet Index, a database of global biodiversity, found that European migratory fish have declined a staggering 93 percent in the past five decades, the highest number on any continent (though large migratory species in Asia, for example, were not included in the report).

That same year, a four-year study spanning thirty-six European countries, in which scientists surveyed almost seventeen hundred miles of river by foot, found that at least 1.2 million obstacles prevent European rivers from flowing freely. That's more than one barrier for every mile of river (or 0.74 barriers per kilometer). The numbers, said Barbara Belletti, a river geomorphologist who led the study at the Polytechnic University of Milan, "show that European rivers are broken."

It's hard to argue against that notion; European rivers have been degraded to the point that many bear no resemblance to their natural state. Yet the damage to most European rivers—exacted on them in the form of dam building, water diversions, pollution, and habitat destruction—was mostly carried out in the relatively distant past, just like it was in the United States. In most of Europe, new dams are no longer being built, and the narrative has in many ways and places shifted—again, as it has in the United States—from focusing on the decline of rivers to what is being done to restore them.

To halt the deterioration of aquatic ecosystems, the European Union in 2000 enacted legislation known as the European Water Framework Directive, calling on member states to achieve "good qualitative and quantitative status" of all water bodies by 2015. The goal may have been very loosely defined, and indeed it was not achieved, with close to half of all EU water bodies covered by the directive adjudged to have not achieved that aim by that date. Nevertheless, it's been considered a very helpful tool for governments, and it enjoys strong support from both conservation groups and much of the public. So it has been renewed with a (perhaps still unrealistic) target of having 100 percent of the EU's freshwater ecosystems in good health by 2027 at the latest.

At the same time, many public agencies and environmental organizations across Europe have in the last couple of decades pivoted toward issues of river and fish health, with transnational networks established to coordinate research and to influence decision-making. For a continent that long viewed rivers as little more than shipping canals and sewers, this has been a striking change. What has been particularly encouraging to see is the groundswell of local activism that has brought with it new and innovative initiatives.

One person at the forefront of this activist movement, and someone I've worked with and come to admire for his enthusiasm and energy, is Herman Wanningen. Growing up in the small town of Dwingeloo in the Netherlands, Wanningen spent a lot of time as a kid playing in a nearby brook, placing rocks and blocks in the stream to alter the flow of water. He was the only one in his family who went on to university studies, earning degrees in environmental chemistry and biology, and later got a job at the local water board as an ecologist.

Like most water management agencies in the low-lying Netherlands, the board focused its work on flood control and ensuring clean drinking water for people. But Wanningen wanted to create healthy waters not just for humans but also for fish. Working on river and lake restoration projects, he took a particular interest in the damaging effect that Holland's many pumping stations—machines used to supply water to canals or drain water from low-lying lands—have on fish. He advocated for policies and technologies that implemented different fishway solutions, in line with the European Water Framework Directive, and organized events that drew thousands of people and garnered considerable media attention.

But Wanningen was not content to work just locally. He started the World Fish Migration Foundation, and he had an idea: what if people around the world came together for a one-day global event to raise awareness around the importance of free-flowing rivers and migratory fish? It was an idea that took off right away, and since its start in 2014 the World Fish Migration Day has grown into a biannual celebration that generates hundreds of events, from fish releases to art contests, involving tens of thousands of people in more than sixty countries worldwide.

The reason, I think, the event has been so successful is that it gives people an opportunity to create and tailor activities in their own places, with that local action connecting to a global issue, so they feel like they're part of something bigger. World Fish Migration is also built around a simple idea: in order to survive, migratory fish need rivers to flow free.

Turning much of his attention to the issue of dam removals, Wanningen founded Dam Removal Europe, a coalition of groups that includes his own foundation, as well as WWF, European Rivers Network, Rewilding Europe, and others. Wanningen says he sees the dam removal movement in the United States as a model for Europe. At the same time, he's aware that conditions there are different. In the United States, the importance of hydropower as an energy source has been steadily waning, and today generates only about 6 percent of electricity usage in the country. In Europe, that figure is 13 percent, with some countries like Norway and Albania getting almost all their electricity from hydropower.

Nevertheless, several large dam removal developments have been initiated around Europe in recent years. For the largest project, work began in 2019 to remove two big dams on the Sélune River in France in order to reconnect the river and the ocean and increase juvenile salmon habitat threefold. In Estonia, the removal of the large Sindi Dam and other barriers along the country's Pärnu River could free up more than 1,850 miles of waterways. But in many places in Europe, taking out large hydroelectric dams that are still vital energy producers is neither politically nor economically viable, and, recognizing this fact, many dam removal advocates have set their sights on smaller targets as well.

The study found that the 1.2 million river barriers in Europe include thousands of large dams. But at least 85 percent of barriers on European rivers are actually smaller structures such as weirs, culverts, fords, sluices, and ramps. These shorter barriers, most under six feet in height, are often overlooked even though their cumulative impact on river connectivity may be substantial. While large storage dams can change how an entire river behaves, almost all barriers, no matter how small, have some effect on river life. For many organisms, it doesn't matter if the barrier is one or twenty meters high; either way they can't get through.

Maybe there is a silver lining in that: mitigation efforts to improve river connectivity may not be as complicated as removing a large dam, as many of these smaller barriers are also obsolete and could be easily taken out. In Sweden, which like most European countries gets almost all of its hydropower-generated electricity from large dams, removal efforts have been targeted at small-scale dams and weirs. A fund recently set up by some of the largest hydropower companies in the country, aimed at improving the environmental standards of all dams, offers landowners and small-scale dam operators financial aid to remove structures.

A similar strategy has been pursued in neighboring Finland. Mitigation efforts there initially focused on removing culverts in small creeks before moving on to target smaller obsolete dams and even small hydropower stations still in use. Meanwhile, a public service campaign on dam removals has been successful in moving opinion in Finland by highlighting the benefits that many Finnish people associate with free-flowing rivers, including various leisure activities, rather than focusing on the subject as an infrastructure matter that people may find abstract.

What this shows, perhaps, is that focusing on what is doable and building from there can often be a winning strategy, especially when you're dealing with such complex issues as river decline, or what some have deemed to be a "wicked problem," a problem that is difficult or even impossible to solve because of its interconnected nature and lack of a single solution. Maybe even the most wicked of problems can be solved, as long as you start somewhere.

There is no doubt we know a lot more today about many of the world's largest freshwater fishes than we did twenty years ago. When I first started working in Cambodia, it was believed the Mekong giant catfish had disappeared from there, but I soon learned that Cambodia was in fact the last refuge for the critically endangered catfish. The first time I saw a giant freshwater stingray, I was blown away not only by its size but by its very existence. Since then, a closely related species in Australia has been discovered. In Mongolia, taimen research has advanced by leaps and bounds, as has our understanding of alligator gars. And the list goes on.

We have learned that the world's giant freshwater fishes are in deeper trouble than we knew. But we also have a much better understanding of the causes of their decline. This is logical, since much of our scientific and conservation work has been focused on figuring out why the largest freshwater fishes are disappearing at such an alarming rate. Sadly, the knowledge we have acquired appears to have come too late for some species, such as the Chinese paddlefish.

One species we know a lot more about now—which is also one of the few megafishes doing well—is Europe's largest freshwater fish, the wels catfish. Native to eastern Europe, the European catfish, as it's also known, was introduced into rivers in southern Europe by recreational anglers in the 1970s and has been thriving there ever since. But only in recent years has the species been properly studied in its introduced range. What scientists have learned about the wels catfish there, and its hunting behavior in particular, is both fascinating and troubling.

One of the leading European catfish researchers is Frédéric Santoul, the fish ecologist from the University of Toulouse who worked with us on the *Monster Fish* episode in France. Like me, Santoul first became interested in the wels catfish after hearing of its pigeon-feasting in the Tarn River. After witnessing that macabre spectacle with his own eyes, he spent a summer documenting the phenomenon. Further research led him to start a European Catfish Working Group with other scientists who were interested in the fish.

They knew the catfish was an extraordinary hunter, and that it could track prey in three dimensions in absolute darkness using its whiskers and lateral line. It would use deep holes, dense vegetation, woody debris, rocks and boulders as resting sites, exhibiting strong habitat preferences even within invaded regions, and it became highly territorial of a preferred location. The catfish were obviously also extremely opportunistic predators, as demonstrated by their feeding on the pigeons. But what the research would come to show is that their opportunism could evolve and hunting strategies change in response to available prey.

One such prey was the Atlantic salmon (*Salmo salar*), one of Europe's commercially most important migratory fish and a species whose populations have, despite conservation and rehabilitation plans, dramatically declined due to the usual human impacts. Owing to their large body size, anadromous adult salmon were historically invulnerable to fish predation during their spawning period migration. But the introduction of the European catfish had changed all that, and, as researchers found out, some wels catfish populations were very good at figuring out when and where to attack the traveling salmon. In France's Garonne River, for example, the catfish would wait inside a fish tunnel to trap and kill salmon migrating through a hydropower plant. The study suggested that at least one-third of the salmon observed at the dam were consumed inside the fishway.

Santoul was amazed at the ability of the catfish to adapt to novel food sources, including Asian clams, which he and his colleagues reported about in one paper. Another study showed how the ravenous wels targeted sea lampreys—primitive, jawless fish that are endangered in Europe—as these creatures moved from the sea into the rivers to spawn. The research showed that at least 80 percent of tagged sea lampreys had been preyed on within one month, and that 50 percent of them were rapidly consumed within days of being tagged.

In yet another study, also in the Garonne River, scientists documented impressive attacks at the river surface of large catfish individuals on allis shad, another commercially important and declining migratory fish species in Europe. The catfish, the researchers concluded, had learned to attack the shad at night while those fish were preoccupied in their courtship displays. An analysis of the stomach contents of more than 250 catfish revealed that shad made up more than 80 percent of their diet. The researchers titled the study "The Giants' Feast."

The studies all reached the same conclusion: the European catfishes were—and are—in contrast to almost all other freshwater megafishes, thriving in their new habitats, and much of it has to do with their adaptive hunting practices. In doing so, they are becoming a serious threat to the important migratory fish, which we know are already in serious decline in Europe. In recent years, Santoul has been sounding the alarm that if stronger conservation plans for migratory fish are not devised by European nations, the cumulative effects of the pressures put on these fish populations, including predation by the catfishes, could lead to their extinction in the near future.

Meanwhile, ecological shifts driven by climate change, including warming temperatures and changes in precipitation patterns, could create even more favorable conditions for the wels catfish to spread. There is evidence that the wels catfishes, which need water temperatures of at least 68 degrees Fahrenheit for their annual spawning, are colonizing colder, previously uninhabited rivers in countries like Belgium and the Netherlands as those water bodies warm. There are also signs that catfish spawning now occurs several times a year in France, instead of just once, as rivers there stay warmer for longer periods.

This leads us to the question: how big could the wels catfish grow as its populations proliferate in the new and welcoming habitats? We know that animals can adapt to environmental change by increasing or decreasing in size over time. We also know that megafishes, like the wels, are indeterminate growers, continuing to grow throughout their life span and getting as big as their environment, diet, and other factors allow.

Individual animals seem to reach a maximum size because the longer that animal lives, the more likely it is to come in contact with predators, diseases, and other threats (like ending up on a fishing hook) that end its life. But what if any or all of those restraints are removed?

What if a species like the wels catfish is free to roam and adapt in an environment that is custom-made for it, allowing it to grow until it finally dies of "old age"? Is it possible that it could grow larger than any other fish in our rivers and lakes? It is an intriguing proposition: as human activities have cut most megafishes down in size, through overharvest and impacts like dams and climate change, could we also be creating the world's largest freshwater fish?

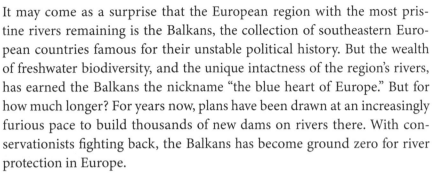

It may come as a surprise that the European region with the most pristine rivers remaining is the Balkans, the collection of southeastern European countries famous for their unstable political history. But the wealth of freshwater biodiversity, and the unique intactness of the region's rivers, has earned the Balkans the nickname "the blue heart of Europe." But for how much longer? For years now, plans have been drawn at an increasingly furious pace to build thousands of new dams on rivers there. With conservationists fighting back, the Balkans has become ground zero for river protection in Europe.

For some time, I'd been talking to my friend John Zablocki, at the time a freshwater biodiversity expert with the Nature Conservancy who has years of experience working in the Balkans, about taking a trip through the region. In the late spring of 2019, we finally made it happen. After flying into the Croatian coastal city of Zadar, I met up with John and his Austrian friend Johannes Schöffmann, who was joining us on a journey through Croatia, Bosnia and Herzegovina, Montenegro, and Albania to take stock of the status of rivers and fish there.

A third-generation baker by trade, Schöffmann is considered one of the world's foremost authorities on brown trout, and had recently finished writing a comprehensive book on trout and salmon, which John later translated from German into English. From the start of the trip, he regaled us with colorful stories from his thirty years of traveling to far-flung locales in search of native trout. In many of those places, he had seen disturbing patterns of plummeting trout populations, and he was justifiably concerned about the trajectory of developments in the Balkans, which is home to a dizzying array of trouts, including the marble trout, the softmouth trout, and the "original" strain of brown trout that has been introduced around the world as a target for recreational anglers.

We stayed one night at the foot of towering karst cliffs just outside Paklenica National Park. From there, we traveled inland to the source of the Una River, which gushes to life from a deep blue hole on a forested mountainside of Croatia, then flows through Bosnia before eventually joining the Sava River and later the Danube at Belgrade, the Serbian capital. Following the river downstream, we soon crossed into Bosnia and entered Una National Park, which is one of only three existing national parks in the country. Known for its rich biodiversity, it is home to 130 bird species and many charismatic animals, such as lynx, wolf, and bear.

The Una River itself is equally biodiverse, with at least thirty fish species, among them the Danube salmon. I was excited about the possibility of seeing one of the largest salmonids on the planet. Despite recorded sizes exceeding five feet, this species was able to live in small, spring-fed rivers like the Una.

The Una River was stunningly beautiful, as were the other deep aquamarine and bright green spring-fed rivers that we encountered in Bosnia, which seemed to appear out of nowhere, flowing over and through porous limestone and creating picture-perfect waterfalls, or swirling over bright sand into azure pools more reminiscent of Tahiti than Europe.

We would stop in various places to snorkel and to fish. The waters were cold, but not unbearably so, and every shade of blue imaginable: deep cobalt, aquamarine, cloudy glacier, and ultra-clear blue that reflected off the surrounding trees and cliffs. It was easy to imagine, in such pristine waters, that there would be equally colorful and uniquely patterned fish. Johannes told us that the dozens of unique strains of brown trout each had their own color and spotting patterns.

To my disappointment, I didn't see any huchen while snorkeling (or many other fish, for that matter), and we weren't catching anything when fishing either. Based on his observations and comparing them to past trips, Johannes said the fish density and diversity had declined. Snorkeling in the past, he had immediately dropped into schools of fish. Those fish were now in ones and twos hidden in dense vegetation or staying in the least disturbed, most inaccessible sites.

As we journeyed through Bosnia and dropped into Montenegro, the natural landscape, dotted with picturesque medieval towns, continued to dazzle. But it was hard to ignore the sense that this land, which had been sustained by its magnificent rivers and withstood the ravages of war for centuries, was about to be transformed. John had shown me maps of hydropower developments throughout the region. On maps showing existing dams, dots were sparse. But the maps showing where dams were planned or under construction were shot through with red dots.

A vision of what the looming hydropower boom might do to one of Europe's last aquatic wonderlands was soon presented to us. Crossing into northern Albania, we passed between tall mountains on one of the steepest roads I've ever been on. As we descended, the road began to run parallel to an emerald green stream, which grew in size and brilliance as we dropped into the valley. The road was treacherous, with blind curves and herds of goats to navigate, but it had been paved relatively recently.

I wondered why such effort and expense had been put into perfectly grading and paving what seemed like a sparsely trafficked road, but I soon got my answer. It led to a newly constructed hydropower project. At the bottom of the valley, the entire river had been diverted from the riverbed into a giant pipe over ten feet in diameter. As we followed the pipe down the valley, it dead-ended unceremoniously at a power station, where the water that the pipe carried was forced through turbines before being dumped back into the hollowed-out river channel. It was an egregious example of economics over environment.

And yet it was not difficult to see why such a monstrosity had been built. For a landlocked nation like Albania, with few fossil fuels or other energy resources, hydropower is a logical way to benefit from a wet, mountainous geography. With a population of less than three million people, Albania, which until it emerged from Communist rule in the early 1990s was long one of the most isolated countries in the world, is still among the poorest nations in Europe. But its economy has been steadily growing, and with it the demand for energy. And so Albania has had to rely on its rivers to produce the power that its people demand.

Most of the country's hydropower is generated from rivers in the north, where there are 170 electricity-producing dams. In contrast, the rivers in the south have remained relatively undeveloped. In fact, one of Albania's main rivers, the Vjosa, flows unimpeded for 120 miles through southern Albania on its way to the Adriatic Sea. That makes it one of the last major European rivers that still runs wild. (At least in Albania; there's a dam built near its origin in the Pindus Mountains across the border in northern Greece.)

The Vjosa has become the flagship for conservation efforts to protect free-flowing rivers throughout the Balkans and beyond. For years, international and local conservation groups, as part of a coalition called Save the Blue Heart of Europe, have been fighting plans by the Albanian government to dam the river, which resembles what the Danube and other central European rivers once looked like in their wild state. The efforts by people like Ulrich Eichelmann of Riverwatch have been successful, with the Albanian government announcing it would scrap all dam plans for the Vjosa and declaring it a nature park.

Not content to stop there, however, the activists have pushed on to demand that the Vjosa and all of its tributaries be turned into a national park, which would make it the first of its kind for a whole river in Europe, and provide far more protection than a nature park does.

On the last day of our trip, I had dinner with John and his friend Velibor Ivanovic, a Montenegrin fishing guide who focuses on catching giant trout. Velibor showed us photos of giant huchen that he had caught on small, hidden stretches of the Danube, various tributaries that had escaped the impacts of industrialization and damming. Many of these stretches, he said, were protected and managed by local fishing clubs. Since we had not managed to see any huchen on our trip, I was relieved to hear that populations of this extraordinary fish appeared to be doing well—and getting big—in these rivers.

Velibor told us about the successes of other fisheries. In Slovenia's Soča River, which, with its emerald green-colored water, is surely also one of the most beautiful rivers in Europe, populations of native marble trout (*Salmo marmoratus*) had recovered. Anglers there were now catching very large fish, with the world-record marble trout weighing in at fifty-five pounds. It was a similar story on the Sava River, the third-longest tributary of the Danube, which runs through several Balkan countries. Velibor said it all came down to river protection, good management, and committed people who can serve as a voice for rivers and fish.

As I left the Balkans, I could not help but think the region was at a crossroads. If the plans for all the hydropower dams were to go ahead, the blue heart of Europe would be reduced to a patchwork of pipes, dams, canals, and power stations. Never mind that few of the projects were economically feasible and were often designed to serve private or political interests. With more than a thousand of the new dams—about one-third of the total—planned in protected areas, the impacts to both terrestrial landscapes and aquatic ecosystems would be disastrous.

But maybe things didn't have to go down that path. With alternative (and truly green) energies, like solar and wind, becoming not only available but cheaper than destructive hydropower for countries like Albania, maybe the leaders of the region, listening to their constituents, would change their minds and make decisions to leave the rivers to flourish untouched. And maybe the Balkans could be a beacon for river protection elsewhere in Europe.

What gave me hope was listening to stories like the ones Velibor told us, about fishing clubs and angling communities coming together to protect and preserve some of the most vulnerable trout populations in the world. My hope sprang from hearing about activists in Albania, where any type of civic protest was totally suppressed for decades, now taking to the streets to voice their support for river protection. It came from hearing about the group of women from Bosnia and Herzegovina who occupied a bridge over the Kruščica River for over five hundred days and nights, while defying violent eviction attempts by the police, to protest the construction of two hydropower plants. I later heard that those women were awarded the Goldman Environmental Prize, the same environmental award that Tsetsegee Munkhbayar had won in Mongolia. It was the second time in three years that activists protesting against hydropower in the Balkans had won the award.

When fish scientists talk about shifting baselines, we usually refer to fish populations. The "baseline" size denotes how abundant a fish population is in its natural state. But there's a problem: that natural state is very difficult to define. Significant changes to a population size may have occurred long before the baseline is set. A diminished population size (caused, for example, by overexploitation of a species) becomes the norm, but it cannot be considered the natural state. Over time, knowledge is lost about the state of the natural world, because we don't perceive the changes that have actually taken place. It's not so much that we forget, but that we never knew.

We can also talk about shifting baselines when it comes to the state of rivers and entire ecosystems. An altered and degraded river becomes the new normal. This is what happens with dams, which may be seen as permanent features. The way the river functions with the dam becomes the standard, or baseline.

I think about this in the context of restoring the Danube River. The baseline of that river's ecosystem and its entire natural functionality has shifted dramatically from the time centuries ago when the Danube spread and ran as it wished, to when industrialization took hold and the river was engineered into something else. That something else has been the Danube's new normal for so long that it's natural to see it as the river's true ecological baseline.

It is not. And yet so often we treat it as such. We see the new normal as something fixed and permanent, something we cannot do anything about, when in reality nothing is permanent, not even the dams of the Danube. Those dams, with their finite life spans, will come down one day. That is a fact. Unless we build new dams, the river will find its natural self again, probably quicker than anyone could ever imagine, and the baseline will have shifted once again.

What we need to do in the meantime is protect what we have and make sure that it does not vanish forever. While the sturgeons of the Danube have disappeared from the upper and middle stretches of the river, they are still found in the five hundred miles of free-flowing river below the dams of the Iron Gates, the stretch that is known as the Lower Danube and that connects the Black Sea to the sturgeons' spawning and feeding grounds upstream. In fact, within the European Union, the Lower Danube is the only river remaining with naturally reproducing sturgeon populations.

In recent years, conservation groups, spearheaded by the World Wildlife Fund, have increased efforts to boost the endangered Danube populations of sturgeons by restocking fish in the river. To mark Danube Day in the summer of 2020, over seven thousand three-month-old beluga sturgeons were released into the river in Bulgaria, following a crowdfunding appeal that drew support from more than one thousand people across Europe.

There are no guarantees that such restocking efforts will be successful—in fact, there are many examples of them not working—and they are not long-term fixes. They are Band-Aids on a problem that requires much stronger measures. But we tend to forget that Band-Aids are also useful. They stop the bleeding.

It may be many years before the Danube River reverses its downward course, but at the same time restoration efforts have been long under way and are producing exceptional results. The Danube Delta, the largest river delta wetland in Europe, has been turned into a protected biosphere that is now one of the finest and wildest places on the continent.

If we focus on what is doable, the day will come when the entirety of the Danube River will enjoy the same status. And who knows, maybe on that day an American exchange student will be wandering around Regensburg and find himself at a picnic table by the Danube, and below giant sturgeons will swim, as if that's how it's always been.

10

Message in a Bottle

Among anglers, the golden mahseer (*Tor putitora*) is one of the world's most prized catches. Not only is it beautiful—golden-hued, with big, reticulated scales—it also has a reputation as one of the world's hardest fighting fish. Author Rudyard Kipling once wrote that the tarpon, another valued game fish, is "as a herring" in comparison to the golden mahseer. But this large carp, which purportedly grows to lengths of up to nine feet, has long been hunted for food by humans and has lost significant portions of its habitat, decimating its numbers throughout its southern Asian range.

Except in one place: the remote Himalayan kingdom of Bhutan. There, where it is known as the tiger of the river, the golden mahseer has benefited—like other endangered species in the country, including actual tigers, white-bellied herons, and golden langur monkeys—from strong environmental protections and the religious reverence bestowed on it.

Nestled in the mountainous fold between India and China, Bhutan is a small kingdom about a tenth of the size of Montana, with a population of less than a million people. For centuries, this "land of the thunder dragon" kept itself cut off from the rest of the world to protect its culture, which is steeped in Buddhist traditions. The country had no roads, no electricity, no motor vehicles, no telephones, and no postal service until the 1960s. A ban on television was lifted only in 1999. It was not until the 1970s that the first foreign tourists were allowed into the country. Even today, the authorities keep a close eye on foreign influences, restricting the number of visitors from outside South Asia by charging them a substantial fee every day to be there, all to keep tourism "low impact" in environmental and cultural terms.

Internationally, Bhutan is perhaps best known for its unique philosophical approach to measuring economic health, through a construct its government calls "gross national happiness," and also for its progressive environmental policies. More than 70 percent of the country is covered in trees—by law, 60 percent must always remain forested—and the kingdom became the first nation in the world to go carbon negative, meaning it absorbs more carbon dioxide than it produces.

Anyone who visits Bhutan, as I did some years ago, is immediately struck by the country's environmental ethos. There are signs everywhere urging people to respect nature. ("Be responsible, don't hurt animals. You will be repaid by good karma.") Mountain climbing is prohibited, because in Bhutan the mountain peaks are considered sacred. As are its rivers. This partly explains why roads are mostly built at high elevations rather than along the valley floors where rivers flow. It also means that what should be a two-hour drive in Bhutan will usually take ten.

There are four major rivers in Bhutan, of which the Manas (or, as it's known locally, the Drangme Chhu) is the biggest. It originates in Tibet and winds its way through Bhutan before joining the mighty Brahmaputra River in India down below. While most Bhutanese rivers are full of fish, including Himalayan trout, mahseers (not just golden but also chocolate mahseer, a smaller but also sparkling species) are found primarily in the Manas and its tributaries in the tropical south.

It has long been forbidden in Bhutan to fish for mahseer, with the golden mahseer considered one of eight auspicious signs in Tibetan Buddhism, representing good luck. In the 1970s, Bhutan's king at the time commanded his guards to protect mahseer spawning areas from poachers, who targeted it as a source of food. In 1995, the mahseer was listed as a protected species under Bhutan's Forest and Nature Conservation Act. Since then, mahseer populations in Bhutan have remained relatively undisturbed, unlike in the rest of its natural range—from Afghanistan in the west to Myanmar in the east—where they have declined by at least half. Today, the species is listed as endangered on the IUCN Red List.

Until recently, there was little scientific knowledge about the golden mahseer in Bhutan, despite its famed status. "It's always been talked about in this almost mythical sense. People hadn't looked at it from a research or empirical perspective," says David Philipp, a fish biologist and chair of the Fisheries Conservation Foundation, a US-based NGO. That changed in 2015 when Bhutan's government, acting on a directive from the current king, Jigme Khesar Namgyel Wangchuck, and working closely with the World Wildlife Fund, recruited Philipp's organization to start a research project designed to learn more about the ecology of the much-revered mahseer.

The researchers, working with a team of Bhutanese forest rangers, set up a series of signal-receiving stations along the Manas River and its tributaries, no easy task in the inaccessible jungle terrain of southern Bhutan. Over two years, they captured, tagged, and released more than sixty golden and forty chocolate mahseers, gathering data about where and how far the fish traveled. They found that golden mahseers migrate longer distances and at greater speeds, sometimes through tough rapids, than anyone had previously thought, and that the fish use warmer, non-snowmelt tributaries for spawning, with individuals returning to the same locations annually. Curiously, the data showed mahseers staying in Bhutanese waters and not moving into the Indian part of Manas National Park, despite being able to do so.

When the findings were presented at an international mahseer conference held in Thimphu, the Bhutanese capital, at the end of 2018, it underscored Bhutan's emergence as the leader in mahseer research and fish telemetry in the region. It also added to the country's growing reputation as a conservation success story.

This is not to say that Bhutan is some kind of environmental Shangri-La (an epithet that is often attached to it), or that it has all the answers. As the country goes through rapid modernization and pursues closer economic integration with the outside world, its natural resources are becoming increasingly vulnerable. With more people moving to urban centers, fewer "stewards of the land" remain in rural areas, where poaching of wildlife has been on the rise. For the migrating golden mahseer, an expansion of the country's hydropower sector, which generates important revenue by selling electricity to India, could pose a major threat. Illegal fishing remains a problem, despite better law enforcement to help stem it.

But what makes me optimistic that a country like Bhutan can ultimately be successful in protecting endangered and extraordinary fish like the golden mahseer is its environment-first mindset. Decisions on development projects, including those involving hydropower, are usually guided by science and environmental impact considerations. When studies of plans for up to four new hydropower plants in Bhutan showed that mahseer migrations would be negatively impacted, the government recommended pausing the construction of those and all hydropower projects in the country. In other instances, hydropower plants have been proposed at higher altitudes where they are out of reach for the migrating fish.

Viewing conservation through the prism of natural capital—the idea that attaching a monetary value to nature will incentivize us to protect it—can be useful. As we've seen with dam removals and other conservation actions, there can be financial benefits from working with nature and turning it into our ally.

At the same time, we cannot regard the natural world only in dollars and cents. That would be to chain it to the same system of financial incentives that is so often destroying it in the first place. There is a great economic argument to be made for protecting big fish, which are often among the most valuable catches. But in the end, it is not the only, or even the principal, reason to do so. We must protect these weird and wonderful creatures because they, like us, depend on clean water and healthy environments to thrive. Most importantly, they, like us, have a right to exist. That is something Bhutan and its people seem to have long understood.

Since I started my work, the prospects for many of the world's largest freshwater fishes have dimmed. The Chinese paddlefish has disappeared, and other giants are poised to follow. Dams are being constructed without an honest attempt to understand and avoid the environmental costs. Invasive species keep spreading unchecked. In many places around the world, habitat loss is accelerating. The cataclysmic effects of our changing climate are no longer a thought experiment for the future; they are a real and present danger.

We cannot ignore these realities, but we do ourselves a disservice if we focus only on the negative. That's not how problems get solved, especially not in the realm of conservation. When problems are presented as irreversible or too big to fix, people lose the motivation to address them. Seemingly small or incremental actions can often change the outlook in dramatic ways for species that once seemed doomed to extinction. After all, there was a time when the alligator gar was considered a trash fish and subject to systematic extermination. But then we realized that gars are in fact nothing like trash, and this change in perception led us to implement simple protective measures that have allowed gar populations to stabilize.

For context and maybe even guidance, we can look to the modern marine conservation movement, which emerged in the 1970s in response to the increasing influence of exploitative human activities, such as overfishing. Few at the time would have put much faith in the future of sharks, which were then likely being killed by the hundreds of millions each year. But shark conservationists worked to change the public perception of these magnificent animals, and eventually their efforts paid off. Today, sharks are a poster child and rallying cry for marine conservation.

For all the horror stories we hear about the state of our oceans—and with the seas mired in plastic pollution and coral reefs dying from acidification, those stories are indisputably true—marine conservation efforts have also produced thousands of success stories, and continue to do so. Some of these stories play out on a limited scale, in the form of beach cleanups or initiatives to hang lights on fishing nets to prevent turtle engagement, while others are larger in scope. Increased awareness of the threats to the oceans, and the need to better manage biodiversity and fisheries, led to many marine reserves being set up around the world starting in the 1970s. Today, nearly 5 percent of the seas worldwide are protected, with almost half of that space made up of no-take reserves where fishing is not allowed at all.

The point here is that marine conservation has had half a century to evolve and develop those efforts. In contrast, the global movement to protect our fresh waters, and the fish that live in them, is still in its early stages. That said, I've seen a remarkable increase in awareness about freshwater issues and especially big fish. When I first became engaged in this work, I would give lectures to school groups, and it was clear that the students had never heard of these issues or seen a giant freshwater fish. Today when I give talks, the kids know the names of all the fish and sometimes scream them the second a photo appears on the screen in front of them. These fish have made it into the public consciousness; now we need to take the next step and protect them.

We are seeing the emergence of a movement of people working together to turn the tide on the freshwater crisis. Collaborative efforts to map out new frameworks for how to tackle the huge problems facing freshwater systems are proliferating. In early 2020, a large group of freshwater scientists presented, in a paper published in *BioScience,* what they called an emergency recovery plan to bend the curve of freshwater biodiversity loss, outlining six priority actions that included accelerating the implementation of environmental flows and improving water quality. The following year, I was a coauthor on a similar paper calling for improving globally coordinated conservation actions.

What we are finding is that these actions have, for the most part, already been implemented successfully in various situations across the globe, providing proof of concept and lessons that can inform how to scale up efforts.

There's not going to be one silver bullet solution to magically repair the damage we've caused to our freshwater systems and the life that depends on them, including our own. To keep rivers productive and free-flowing, with fish populations healthy and robust, we must take action where we can.

It's important to not stare ourselves blind at the big picture, which can be overwhelming. Because if we do, we may be ignoring existing solutions that have been staring back at us all along. To put it another way, we might miss the trees for the forest.

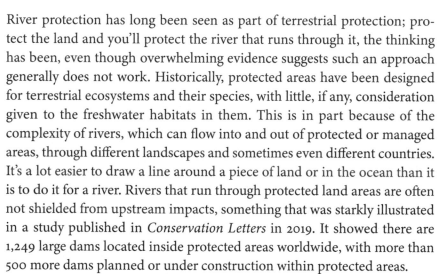

River protection has long been seen as part of terrestrial protection; protect the land and you'll protect the river that runs through it, the thinking has been, even though overwhelming evidence suggests such an approach generally does not work. Historically, protected areas have been designed for terrestrial ecosystems and their species, with little, if any, consideration given to the freshwater habitats in them. This is in part because of the complexity of rivers, which can flow into and out of protected or managed areas, through different landscapes and sometimes even different countries. It's a lot easier to draw a line around a piece of land or in the ocean than it is to do it for a river. Rivers that run through protected land areas are often not shielded from upstream impacts, something that was starkly illustrated in a study published in *Conservation Letters* in 2019. It showed there are 1,249 large dams located inside protected areas worldwide, with more than 500 more dams planned or under construction within protected areas.

Now, people are increasingly thinking of river protection in new and innovative ways. One movement, for example, is focused on providing legal protection for rivers. In 2017, New Zealand became the first country to grant a river (in this case the Whanganui River) legal rights the same as those of humans, meaning that in a court of law the rivers are treated like living entities. Since then, Bangladesh has done the same with all of its rivers, while the city of Toledo, Ohio, passed what's known as the Lake Erie Bill of Rights to protect its shores, making it one of several cities in the United States to pass legislation recognizing the rights of nature.

When I was a graduate student, my academic advisor, Peter Moyle, was part of a lawsuit to restore flows to our town's creek, which was poorly managed by the local water authority. California law stated that creeks and rivers must be kept in "good condition," which was interpreted as healthy enough to sustain native fish (a logical definition, in my opinion). The legal action was successful: water flows were restored, and salmon returned to the creek after a long absence.

Another form of river protection, and one that has long intrigued me, centers around fishing reserves. These are protected areas established in a river system in which fishing is restricted or not allowed. The efficacy of such no-fishing zones has long been demonstrated in marine environments, but far less attention has been given to them in freshwater habitats. They are, however, common in Southeast Asia, where they are often established as sacred pools around religious temples.

There was one particularly interesting grassroots network of fishing reserves in northwestern Thailand, which I was curious to see for myself. So, in the spring of 2019, Stefan and I flew to Chiang Mai, where we were going to spend the night before continuing by car to a river called Ngao, which is where the reserves had been set up.

It was nice to be back in Chiang Mai, the place where I had studied for my Fulbright scholarship. More than two decades had passed since I first arrived there, and much had changed in the city. As a Fulbright scholar, I had stayed in a small, secluded house near a forest temple. Now the same neighborhood was bustling with trendy hotels and a Michelin-starred restaurant.

The following morning we rented a car and set off on the journey west, into the mountainous hinterlands of western Thailand, through which the Ngao River runs for forty miles as part of the Salween River basin in neighboring Myanmar. The car trip took longer than expected, with the roads becoming increasingly narrow, and it was late afternoon when we finally arrived in Ban Luiy, one of the many villages dotting the hillsides of the meandering Ngao River valley. There, we were met by Aaron Koning, an American aquatic ecologist who had been studying the reserves for seven years. I was told he had made some remarkable discoveries.

Aaron brought us to a viewing platform that overlooked the river. Down below, scores of blue mahseer, a smaller type of mahseer than the ones found in Bhutan, congregated in crystal clear waters while a group of children splashed nearby. Aaron explained that this stretch of the river, extending about a thousand yards, had been designated as a fish sanctuary twenty-five years earlier.

At that time, people in Ban Luiy had noticed that their fish catches in the river were declining and the fish they were catching were getting smaller. At the urging of a local NGO, the villagers decided on a radical solution: they would set aside a stretch of the river below their village and declare it strictly off-limits to fishing. In the years after that, other villages had done the same thing, presumably because of the positive results the experiment at Ban Luiy had yielded.

But could such reserves, covering only short stretches of river and lacking physical boundaries, really have an impact on fish, and especially migrating fish that may travel up and down the entire river system? That was the question Aaron had set out to answer when he came to the area in 2012 to begin his study.

First he had to find out how many reserves there actually were. No one seemed to know. Traveling upstream, Aaron was amazed to discover that almost every single community along the way had set up its own fish reserve, with around fifty in total now operating on the main river and its tributaries. Selecting twenty-three of the reserves to study in depth, Aaron spent months at a time over the next seven years living in area communities, all the while interviewing villagers about their fishing, and snorkeling the waters inside and outside the reserves—some of which were no bigger than a kiddie pool—to count and measure fish.

What he discovered was astonishing: virtually all of the more than fifty species of fish found in the Ngao River system were using the protection the reserves afforded, and, in doing so, they were able to substantially boost their numbers. The study showed that, compared to nonprotected sections of the river, the reserves enjoyed more than twice the total number of fish and over twenty times the total weight of fish. Of particular interest to me was that the greatest conservation benefits were enjoyed by the biggest fish, which in the Ngao River include large barb and other species. These fish were found almost exclusively within the protected areas.

The research showed, perhaps not surprisingly, that older and bigger reserves were more successful, because they offered more time and space, including more kinds of habitat, for fish populations to rebuild and for rare species to reestablish. But even reserves set up recently showed clear benefits from being spared intense fishing pressure. Reserves placed closer to each other saw more positive effects than others, and those located closer to a village tended to have an advantage, most likely because villagers were better able to enforce the rules.

Equally important, the reserves also provided strong benefits from a fisheries point of view, with catches outside of the reserves, where the villagers were allowed to fish, showing significant increases. This, Aaron told us, was evidence of a so-called spillover effect that occurs over time; reserves become crowded and fish start to leave, allowing fishers to catch the fish in nonprotected areas. These were significant findings. It was the first time anyone had so clearly and quantitatively demonstrated the efficacy of freshwater fishing reserves in the region and shown their benefits from both a conservation and fisheries standpoint.

What particularly impressed me was how organically these fishing reserves had been established and were now being run. It was a true grassroots effort, without even local government involvement. These villagers had decided on their own that this was something they could do to protect the one resource they cared most deeply about. The job of guarding the reserves fell to the villagers themselves, which in a way made the job easier. Instead of having a few park rangers patrolling forty miles of river, there were two thousand people who lived there and could do it every day.

We spent the following days moving around the area and talking to communities. It was easy to see the pride that people there felt about their efforts to protect the river and its fish. It was clear that those efforts had created a great sense of harmony and purpose among the villagers. Pannee Phoemchatchai, a thirty-eight-year-old seamstress in Ban Luiy, even talked about managing the village fish reserve as a divine responsibility. "God created the fish and told humans to look after them," she said, adding that when she felt down, she could always go to the river and look at the beautiful fish. "It eases my mind," she said with a smile.

In most of the world, the decimation of large freshwater fish populations has occurred in recent decades. One exception is North America. There, many of the largest freshwater fishes saw their numbers start to dwindle more than a century ago, with many species falling victim to the one-two punch of wanton harvest and habitat loss through dam building. Some of these giants, like the Colorado pikeminnow, have never recovered. But there are other species whose futures once looked doomed that have recently bounced back, at least in some of their native habitats.

We've talked about the alligator gar as an example of such a species. There is also the Mississippi paddlefish (*Polyodon spathula*), also known as the American paddlefish. One of the largest North American freshwater fish species, the Mississippi paddlefish can grow over seven feet long and attain weights of up to two hundred pounds, according to FishBase. With the Chinese paddlefish declared extinct, it is the only remaining paddlefish species in the world.

There was a time when the Mississippi paddlefish could have gone the way of its Chinese cousin. This ancient and strange megafish, with its sharklike body and prominent, paddle-shaped snout, was once abundant throughout the Mississippi River basin, its core habitat, with populations extending into the Great Lakes and as far east as New York and Maryland. But overfishing (much of it in the form of poaching, with the paddlefish highly valued for its caviar), habitat destruction, and pollution dating back to the late 1800s decimated paddlefish numbers, with the species becoming extirpated from its peripheral range.

Before the decline had reached a point of no return, however, the Mississippi paddlefish was afforded various protection measures under state and federal laws. This stabilized populations, albeit at much lower levels. Since then, federal and state hatcheries have been producing paddlefish for stock rehabilitation programs, with the most prominent implemented in Pennsylvania, where the paddlefish has been reintroduced in the upper Ohio and lower Allegheny Rivers. The fish has also benefited from pockets of intact habitat in the upper Missouri and Yellowstone Rivers, where self-sustaining populations have always existed and still thrive. The paddlefish has also been afforded international protection under the Convention on International Trade in Endangered Species of Wild Fauna and Flora (CITES).

Today, the Mississippi paddlefish is found in almost two dozen US states, with about half allowing it to be fished in-state for sport. With reduced distribution and numbers, as well as a mix of wild and hatchery-supported stocks, the status of the species is a far cry from what it was two centuries ago. But compared to many endangered freshwater megafish, the paddlefish is doing relatively well, even though it's still listed as vulnerable on the IUCN Red List and though most populations are not self-sustaining but rely on restocking for their survival.

There is a North American relative of the paddlefish, a species that shares much of its troubled history, that I believe has more to teach us about sustainable management and recovery of freshwater giants: the lake sturgeon.

Also known as the rock sturgeon, the lake sturgeon (*Acipenser fulvescens*) is one of the few true freshwater sturgeons in the world. It is also one of North America's largest freshwater fishes, reportedly capable of reaching up to nine feet in length and 275 pounds in weight. With its sleek shape and rows of bony plates on the sides of the body, it has that cool sturgeon look, a bit like an armored torpedo. It uses the two sets of whiskerlike barbels near its mouth to locate bottom-dwelling prey, such as snails, mussels, clams, crayfish, insect larvae, and fish eggs.

Despite the name, lake sturgeons are also found in rivers and once roamed watersheds from the Mississippi River in the west to Hudson Bay in the east. They are said to have been so abundant (and large) about two centuries ago that they made up an estimated 90 percent of the biomass in the Great Lakes. But they came to suffer the same, or possibly even worse, fate at the hands of humans as the paddlefish.

Accounts from the mid-1800s tell of lake sturgeons being killed en masse as a nuisance fish, with carcasses piled up on the shores to dry and be burned, fed to pigs, or dug into the earth as fertilizer. Things got worse when the value of the lake sturgeon's meat and eggs was realized and commercial fisheries began to target the species. In the late 1800s, over five million pounds of lake sturgeon were reportedly taken from Lake Erie in a single year. Pollution and dam construction made matters even worse, and by the early 1900s the fishery for it had largely collapsed. It is estimated that lake sturgeons have declined by more than 99 percent from their historic numbers, and the species is today listed as threatened or endangered in nineteen out of the twenty US states that are within the fish's original range.

So why would the lake sturgeon be considered a success story? Because there is one state in which its population is thriving, thanks to more than a century's worth of dedicated conservation efforts and smart, adaptive management (that is, management based on good science and data). That state is Wisconsin, home to fifteen thousand lakes and almost as many miles of navigable streams and rivers. In Lake Winnebago, the largest lake entirely within the state, and the watersheds around it, which include its two primary tributaries, the Wolf and Fox Rivers, the lake sturgeon has found a sanctuary that is helping to bring the species back to health.

Back in the day, lake sturgeons were slaughtered in Wisconsin as they were elsewhere. The difference is that officials in the state stepped in early to stop that indiscriminate killing. This led to a statewide moratorium on lake sturgeon fishing in 1915, designed to give the species breathing room to stabilize and recover. When lake sturgeon fishing was allowed to resume for recreational purposes in the state in 1932, it was to be closely regulated and monitored.

Not long after, biological studies of the lake sturgeon began in earnest, with the first harvest assessment made in 1941. In the decades to follow, state biologists tirelessly conducted studies on the age, growth, and diet of the lake sturgeon, building a comprehensive understanding of its life history (which includes the mind-blowing fact that female lake sturgeons may live to be 150 years old). Along with assessments of spawning fish populations on rivers like the Wolf, this information was used to determine that 5 percent of the harvestable stock of lake sturgeon could be taken in any year in order to sustain the population.

This quota is used today to manage an annual sturgeon spearing season, which is restricted to a few weeks in February. It is a very popular event, with anglers flocking by the thousands to frozen Lake Winnebago and the so-called Upper Rivers to take part in a ritual rooted in the traditions of the Menominee and other Wisconsin Indians. With catch numbers so strictly capped, only a minority of anglers will land a fish before the quota is reached. But those who do may catch a monster. In 2012, one angler speared a 7-foot 3-inch, 240-pound female sturgeon on Lake Winnebago. Estimated to be 125 years old, the fish was tagged and released by scientists from the Wisconsin Department of Natural Resources.

Then, approximately two months after the spearing season, thousands of more visitors are again drawn to the area, or, more specifically, the Wolf River, to witness another annual occurrence: the sturgeon spawning. Over a few days onlookers have the opportunity to watch these ancient fish up close as the sturgeons cut up and cavort in the shallow watersheds. A force of volunteers, calling themselves the Sturgeon Guard, will dispatch along the river to protect the fish from poachers. State officials, meanwhile, have for many years been using this window of time to catch sturgeons and tag them to gather more data about the fish. The information gained from these survey efforts proves crucial for sturgeon management.

As a result of this work, the Lake Winnebago system now has the largest self-sustaining stocks of lake sturgeon in North America. It is also helping to revive the species in other parts of the country, with Lake Winnebago providing brood stock for an artificial propagation program. It shows what science-driven management, community support, and a long-term commitment to the preservation of large freshwater fish can accomplish. And that is worth calling a success.

A decade of filming more than thirty episodes of *Monster Fish* has taken me to several truly special locations around the world. In Kamchatka, in Russia's far east, I hiked through lush volcanic forests to reach deep-blue spring-fed pools filled with spawning salmon. In the tropical paradise of the Solomon Islands, I swam in pristine streams along with schools of giant eels. During several expeditions to Australia, I encountered an astounding diversity of large aquatic predators in some of the starkest and most spectacular landscapes on the planet.

One of my absolute favorite filming locations is Guyana, a nation about the size of Idaho situated on the northern coast of South America. Eighty-five percent of Guyana's landmass is covered in tropical rainforest, and the country is one of the world's great biodiversity hot spots. The south-central region, known as the Rupununi, is particularly pristine, with more than 1.5 million acres of protected land and the Amazon and Essequibo river systems teeming with most of Guyana's close to five hundred freshwater fish species.

My first experience in Guyana was in early 2014 to film the first episode of the show's fifth season. In Georgetown, the Guyanan capital, I met up with the crew, which included producer Erin Buxton and cameraman Rob Taylor, as well as Duane de Freitas, a resident fishing guide who was going to work with us on the almost two-week-long shoot.

From Georgetown, we took a bush plane and headed deep into the Rupununi, landing at a dirt airstrip cut out of the jungle. As we landed, our plane was immediately met by an excited group of people from the indigenous Wai Wai community. They seemed eager and happy to see us, at least partly because we were delivering food, mail, medicine, and other supplies from the capital. Among the items we brought was a full-size gas oven. I had assumed this was for the village bakery, so I was surprised when, the following morning, I saw it standing in one of our small expedition boats. Apparently someone had made a deal to borrow it for our trip. It meant we would have fresh-baked bread in the jungle for the next week and a half.

For the show, we usually set out to build each episode around one species of fish. In Guyana, the plan was to fish for the *Hoplias aimara,* also known as the wolf fish. While it is the largest of thirteen species in the *Hoplias* genus, this predatory fish is not by my definition a megafish, growing only up to four feet long and eighty pounds. But it is a monster, with inch-long teeth and a bite from which it will not let go. It doesn't usually bite humans, but I had been told that local fishers feared it more than piranhas. (One might say piranhas are the chihuahuas of the aquatic world, while wolf fish have the build and bite of a pit bull.)

One of the reasons we had chosen the wolf fish as our target fish was that we had been virtually guaranteed that we would catch one. But once we got on the Essequibo River, Guyana's largest, it quickly became clear that catching fish—wolf fish or any other species—was not going to be a problem. The river was chock-full of all kinds of amazing fish: tiger fish, zip fish, peacock bass, electric eels, many kinds of catfish, piranhas, payara, and, yes, wolf fish. Stingrays and freshwater turtles were so numerous that the river bottom seemed to move at times as their dark shadows glided through the shallows.

One day, Duane and I were fishing off the same side of a small canoe when—within seconds of each other—we hooked into separate fish. Both fish seemed big, and as we struggled to reel them in their zigzagging resulted in us tangling our lines. Ultimately, we were able to bring the fish up. To our surprise, we had both caught good-sized catfishes but of different species. (Mine was a surubim, the long-whiskered catfish that I caught during filming in the Brazilian Amazon.) It drove home to me how diverse and healthy the Essequibo River system was.

I was intrigued by the aggressive looking weapons—teeth! spines! shocks!—and feeding behavior of all the predacious fish. The piranhas, which were everywhere, would take chomps out of their prey with their powerful jaws, similar to the way we bite into an apple. The wolf fish would grab onto its victim with conical, almost canine teeth, twisting and ripping off chunks. The electric eels would come out of the tree branches and roots on the river's edge and drift motionless toward their prey before zapping them with six hundred volts.

These fish were not typically dangerous to humans, but the same could not be said for the stingrays. Although timid by nature, the rays often caused injuries by stinging people who inadvertently stepped on them. This happened one day to our boat driver, leaving him in considerable pain. He told us it was the seventeenth time he had been stung. You had to be careful on land, too, as I found out when one of the guides discovered a small but very poisonous green viper under my tent. (This explained why most people in our group chose to sleep above ground, in hammocks.)

Moving down the river, we saw wildlife everywhere: caimans and giant river otters; enormous anacondas and other snakes swimming across the river; small nests in the sand that had been created by giant freshwater turtles, one marked with blood and the tracks of an attacking jaguar. We followed a huge tapir as it struggled to get across the river, its large snout pointed out of the water for air. From high above us, in the canopies of towering trees, monkeys and parrots called and cawed.

For the most part, we didn't see any people, and there was a reason for that. The whole of Rupununi, an area three times the size of New Jersey, had only twenty-five thousand inhabitants. It was clear that the indigenous communities who did call the area home lived lightly in their surroundings, without leaving the deep scars that we humans so often inflict on the natural world around us.

The trip was a success, and back at National Geographic everyone was surprised and excited about the footage we had produced. So excited, in fact, that discussions soon began about getting back to Guyana to film another episode. It seemed like we had found the perfect location for the show. Not only was Guyana visually arresting, with pristine rivers and seldom seen wildlife, we were also guaranteed to catch fish, which was an important consideration given the time and resources that went into making the show.

And so a year later I found myself back in Georgetown to film another episode from Guyana for the following season of *Monster Fish*. We were returning to the same region, the Rupununi. But this time we were joined by a different fishing guide, Ashley Holland, and we planned to build the show around a different fish, the payara, also known as the vampire fish (*Hydrolycus scomberoides*). Like the wolf fish, the payara is not a megafish—it has a max length of under four feet—but it too has a fearsome look, with large eyes and inch-long fangs protruding from its lower jaw, which it uses to skewer its prey.

From Georgetown, we flew to Apoteri, a small jungle village where we had spent some time on the first trip, and from there we traveled up the Rewa River, a tributary of the Essequibo. The fishing was even better this time around. From the start, we were catching a lot of fish of all varieties. Much of that success, I felt, had to do with Ashley, who turned out to be an excellent guide.

Born in Britain, Ashley had spent more than twenty years in Guyana, where he now ran an ecotourism operation in partnership with indigenous communities. Laid back in style, he was one of those people clearly most comfortable in the bush, having led more than a hundred expeditions into some of the most hard-to-reach places in the world. He was knowledgeable and compassionate, truly caring about Guyana and its people. I also appreciated that, when it came to fishing, he was like me, less focused on the perfect angling technique and more curious about the fish and in it for the adventure.

During the day, we piled into narrow canoes powered by small outboard motors, fishing deep pools (best for catfish), submerged logs and trees (best for wolf fish), and rapids and waterfalls (the favorite habitat of vampire fish). At sunset, I usually set up some fishing rods on the beach near the camp and would sit there before dinner or after eating and before going to bed. Most of the fish I caught that way were redtail catfish (*Phractocephalus hemioliopterus*), a beautiful species with a bone-hard head, sharp white stripe along its body, and striking red tail.

One night we took the boat out to try our luck in a stretch of rocky reefs and deep pools just above our camp, a site where Ashley had told us he had caught a big lau lau (*Brachyplatystoma cf. vaillanti*), also known as a goliath catfish. I was fishing in deep water, down toward the bottom of the river, when suddenly I got a bite from something that felt big. For the next fifteen minutes I struggled to reel it in. When I finally managed to bring the fish to the side of the boat, and Rob the cameraman got some lights on it, I could see that it had long whiskers. It was a lau lau, and it was big enough to give me pause. At almost five feet long, it was the biggest goliath catfish I had ever seen, let alone caught.

I was able to get into the water with it, on the edge of the deep pool, in the middle of the night and in the middle of the Guyanese wilderness. As I slipped into the river, I immediately felt more in tune with the fish. I was in its watery world and could sense the immensity of the river around me. A few minutes later I let the fish go, and it moved gently away from me and down to deeper water.

There were, as we had expected, plenty of payaras in the river. They were camera-ready fish: scary-looking, with gaping mouths revealing giant fangs and huge tongues, but also sleek, with slate gray backs, white bellies, and winglike fins. At one point I caught a large payara in a wild stretch of rapids far upriver, in an area we knew was home to caimans, anacondas, and piranhas.

For the release of the fish, I dove down with it. As the water darkened below me, I let go of the fish. But just as I did, it turned back toward me in a reflex, as if to attack. I didn't think it would bite—and it didn't—but it did occur to me that I was in the water with a lot of things that could. However, nothing ever did bite. Even in such a wild place, I was left feeling like the animals there were much more scared of us than we were of them.

I left Guyana in awe of the place. It was inspiring to know that there were still areas in the world where giant fish ruled and the rivers remained wild. Like Bhutan, Guyana had made impressive commitments to protect its natural riches through sound legislation and good conservation policies. Its lush inland forests acted as a carbon sink, with the country becoming one of the few in the world to attain carbon neutrality, capturing at least as much greenhouse gas emissions as it released. It had achieved this, in part, due to a groundbreaking deal struck with Norway, an oil-rich country seeking to offset its own emissions, with the Norwegians paying Guyana to conserve forests. It was a potential model for other countries in the effort to monetize the battle against climate change.

And yet I couldn't help but worry about the future of Guyana's extraordinary wilderness and its aquatic biodiversity. Like Bhutan, Guyana was a biologically rich but financially strapped country. Would its environmental commitment remain firm in the face of economic hardship? Or would it, like so many other countries, be seduced into pursuing short-term economic gains at nature's expense?

The Chinese government and corporations had increased investments in Guyana's transportation sector to support shipments of a growing supply of Brazilian soybeans. Gold mining was expanding, with open pit mines and mercury use threatening the health of the forests and rivers. Where would the next roads be built for the expansion that was happening? Would there come a time when the wondrous waterways that we had traveled in the Rupununi were marked by mines or turned into shipping canals, devouring the wolves and vampires living within them?

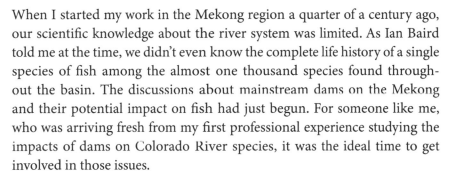

When I started my work in the Mekong region a quarter of a century ago, our scientific knowledge about the river system was limited. As Ian Baird told me at the time, we didn't even know the complete life history of a single species of fish among the almost one thousand species found throughout the basin. The discussions about mainstream dams on the Mekong and their potential impact on fish had just begun. For someone like me, who was arriving fresh from my first professional experience studying the impacts of dams on Colorado River species, it was the ideal time to get involved in those issues.

So much seemed to be at stake. While the impacts on the Colorado River had occurred many decades earlier, similar impacts on the Mekong were yet to be truly felt. This lent urgency to my research, especially since there were huge consequences to what was happening, given that tens of millions of people depended on the river for their food and livelihoods. It was something I was constantly reminded of when, for example, I would be out on the river and see a group of kids, each carrying a small bundle of fish.

Ultimately, it was the giant fish that really captured my imagination. Before I arrived in the Mekong, I had known nothing about these leviathans: giant catfishes as big as grizzly bears; stingrays that required ten men to lift them; and the giant barbs with scales as big as the palm of my hand. They had seemed like local legends until I encountered one enormous species after another.

There was a glaring need to learn more about these giant fish and the river system they inhabited. So I was happy when, after finishing my PhD, I was offered a part-time job as a consultant on a project called the Mekong Wetlands Biodiversity Program. Run by the International Union for Conservation of Nature (IUCN) and the Mekong River Commission, the program focused on biodiversity issues, which made it stand out in a region where most Mekong River research involved commercial fisheries. Four flagship species were selected to be at the forefront of the conservation work: the Irrawaddy dolphin, sarus crane, Siamese crocodile, and the Mekong giant catfish. I was put in charge of all activities related to the Mekong giant catfish.

I immediately developed a conservation action plan that involved tagging fish in northern Thailand to identify their spawning grounds, monitoring catch in Cambodia, and gathering a team of catfish experts to plan out next steps. I poured all of my energy into the job, buoyed by the sense that positive change might be coming for the giant catfish, with people finally willing to commit the resources needed to protect the species.

But after two years, funding for giant catfish activities dried up and our plans were abandoned. My hopes turned to resignation. It seemed clear that biodiversity conservation in the Mekong would forever play second fiddle to competing and far better funded interests like hydropower.

Soon after this, I began filming *Monster Fish,* and before long the television show was taking up most of my time. While traveling around the world for filming, I continued to visit the Mekong region. We did several shows on various fish in Southeast Asia, including stingrays, giant barb, wallago, and snakehead. But after the pilot episode, we never did another show on the Mekong giant catfish. There was a simple reason for that: finding giant catfish to film was almost impossible.

In fact, for several years I didn't hear of any Mekong giant catfish being caught. A report showed that a serious decline in the harvest of the fish had occurred between 2000 and 2010. Catches in Thailand actually dropped to zero, and fishing for the giant catfish was outlawed. In Cambodia, Mekong giant catfish stopped being reported from the dai fishery around 2008, and the status and prognosis for the species in that country was unclear.

Fish that were caught evaded official statistics by being exported illegally to Vietnam, where giant catfish was on the menu in luxury restaurants. Still, the most obvious explanation for the drop in recorded catches was, of course, that the species population had plummeted and was now in crisis. The many pressures on the Mekong, most notably from overharvest and hydropower dams, seemed to have driven the giant catfish to the brink of extinction.

Then, in late 2015, when Stefan and I met up in Phnom Penh, the Cambodian capital, to research some stories, Phanara, my old colleague, told me that a Mekong giant catfish had been caught at the dai fishery north of the city. When we arrived at the dai, a throng of people had gathered on the house float and on the shore. It included the director-general of the Cambodian Fisheries Administration and a news crew from Cambodian Television. This was a big deal. As Phanara and I got into the murky river, the fish was pulled up and placed on a tarp held in the water. It was far from the biggest Mekong catfish I had ever seen, but it was still nearly seven feet long and probably weighed upwards of 250 pounds.

Most importantly, it seemed to be in fairly good shape, with only some light scraping as a result of being caught in the net. We set to work attaching a tag at the base of the fish's dorsal fin. Although we were no longer running any organized tagging operation, it was still important to tag this rare animal so that we could get information about it in case it was recaptured.

After tagging was completed, the director-general performed a brief blessing, which included spraying perfume on the fish. Then it was time for Phanara and I to do what we had done so many times before: move the giant catfish into the middle of the river to release it. Once there, I jumped into the water, cradling the fish in my arms before diving down with it and letting it go. Back on shore, I couldn't help but wonder if it would be the last time I went through this ritual.

A few months before that trip to Cambodia, Rob Lee, Nat Geo's director for science, conservation, and exploration told me that National Geographic had been in talks with USAID, the US Agency for International Development, about a research and capacity-building program focused on the Lower Mekong River basin. Rob asked if I would be interested in participating in the development of the project.

It was a dream opportunity for me: a project focused on aquatic biodiversity in the Mekong, with the financial support and long-term commitment of the US government, run in collaboration with National Geographic. It was perfect.

As the discussions progressed, however, it turned out that National Geographic, for various reasons, could not be the institutional home for the project. But there was another possibility: the project could be run through

the University of Nevada, Reno, where I had been on the faculty for over
a decade. The university already had a good working relationship with
USAID, since my friend and colleague Sudeep Chandra, who was heading
the UNR Global Water Center, had just implemented a similar USAID project in Guatemala.

Sudeep and I—with the help of many partners—developed a proposal,
and the project was formally approved in the fall of 2016. It launched in
early 2017 under the name Wonders of the Mekong. Based in Cambodia, its
mission was "to maintain the natural resources and integrity of one of the
most important river systems in the world." This would be accomplished
through a cooperative agreement with USAID, utilizing a combination of
research, training, and outreach. National Geographic would support us on
the educational and storytelling side of things.

Sudeep and I were named co-principal investigators, and we set out to
recruit a team of experts and staff from both Cambodia and the United
States. I wasn't sure we'd find qualified Cambodians with strong scientific
credentials in river research, but my worries proved misplaced. Many of
the young people who had started out in their careers at the Mekong River
Commission when I was doing my PhD research two decades prior had
recently earned their doctorates, and some would go on to form the backbone of our Cambodian team.

I liked the name "Wonders of the Mekong." It spoke to the magic of
the river and its extraordinary bounty. It had a sense of optimism about
it, which was crucial. Because the dominant narrative seemed to be that
the Mekong was dying and nothing could be done to save it. That stance
didn't sit right with me. Not only did it discount the immense importance
of the river to people and wildlife, but it was also defeatist. How could we
work together to safeguard the river if we had already determined it was
lost? And it wasn't like the river had completely stopped functioning. It was
still delivering the riches that had sustained both people and animals for
thousands of years.

At the same time, I was clear-eyed about the enormous pressures bear-
ing down on the Mekong. Chinese dams in the upper reaches of the river,
from which the Chinese authorities withheld or released water without
regard for what happened downstream, were wreaking havoc on the river's
natural rhythm, with disastrous results for fishers and wildlife alike.

Laos had gone on a hydropower building spree in the Mekong's tributaries and was finally nearing the completion of two big dams on the river's
main stem. These were the proposed dams whose potential impacts on

migrating fish I had come to study as a Fulbright student. There was little reason to believe those impacts would be anything but calamitous.

Of great concern was the deteriorating situation on the Tonle Sap Lake, where fish catches had been shrinking for years. Much of the flooded forest that ringed the lake—vital feeding grounds for hundreds of fish species—was being turned into farmland. Massive forest fires in 2016, exacerbated by drought and climate change, had wiped out as much as one-third of the flooded forest.

In the Mekong delta in Vietnam, riverbeds were eroding because of rampant sand mining and groundwater extraction, leading to the intrusion of seawater that increasingly threatened the most productive rice fields in the world. As a result of plastic pollution, the Mekong was now transporting more than forty thousand tons of plastic garbage into the ocean each year. And looming above it all was the escalating threat of a changing climate.

Higher temperatures cause more frequent algal blooms and reduce dissolved oxygen levels in freshwater systems, which can cause serious fish kills. Climate change is also leading to more frequent droughts because of shifting precipitation patterns. This is particularly true in the Mekong region, where the monsoon seasons are becoming shorter and more unpredictable.

This was vividly demonstrated in 2019, when the monsoon rains failed to arrive. Dry conditions, driven by the El Niño weather phenomenon—the warm, wet half of a naturally occurring weather cycle—persisted well into July, at which point the Mekong water levels dropped to their lowest in more than a hundred years of recorded history, a situation made worse by dam operators upstream withholding water for their own purposes.

Soon, strange things began to happen. In some places in the north, the Mekong slowed almost to a trickle. Farther south, the usually chocolate-colored water began to turn a brilliant blue, a sign that the river had been stripped of much of the sediment it normally transports and which enriches the soils of the basin. Known as "hungry water," such conditions are highly destructive, as the water eats away at riverbanks and causes erosion.

Fish catches from throughout the system, but in particular the Tonle Sap Lake, dwindled further, with the fish being caught often so emaciated they could only be used to feed other fish. When the rains were delayed again in 2020, the flood pulse driving the entire system failed to fully materialize. More and more, people were talking about the Mekong River having reached some kind of "tipping point" from which it would not be able to recover. But for all the troubling developments in the Mekong, there were also bright spots.

In late 2019, China, typically secretive about water management data, made a deal with the Mekong River Commission, to which it had previously paid little attention, promising to share river information and to notify neighboring countries of any unusual dam operations. Soon thereafter, the US State Department and the Stimson Center launched the Mekong Dam Monitor, a tracking tool that uses remote sensing and satellite imagery to provide real-time reporting on climate conditions, estimated river flows, and dam operations.

With solar technologies surpassing hydropower in terms of both price and effectiveness, the Asian Development Bank dramatically increased funding of solar projects in the region. So-called microgrids, which operate autonomously from traditional electricity grids, were becoming increasingly instrumental in bringing power to rural areas.

Community-managed protected areas, where fishing was restricted or banned, were becoming more common throughout the Lower Mekong. About 15 percent of the Tonle Sap Lake had been designated as no-harvest reserves, with that number to be increased in the future. Science in the region continued to expand, with new research centers, monitoring programs, and technology being implemented. These were all important bright spots: the beginning of what would be needed to turn the tide for a river that had run astray, but whose course could still be corrected.

~~~

When the COVID-19 pandemic hit, Cambodia implemented strict border restrictions that made traveling into the country very difficult. Miraculously, Cambodia did not record a single death during the first year of the pandemic, but the tough restrictions, which included a fourteen-day quarantine requirement for all visitors, were kept in place through much of 2021. The US State Department advised against all travel to Cambodia. This meant that most of the US-based members of the Wonders of the Mekong team, including myself, were not able to travel to the country for almost two years.

But it did not mean a slowdown of the project. Our teams organized training seminars online that reached more than a thousand people a year, on topics ranging from fish tagging and catch monitoring to science communication and photo and video storytelling techniques. As the project adapted, more responsibilities were placed on our team in Cambodia, which stepped up and did an excellent job.

Much of the field research focused on fisheries, in particular in the Tonle

Sap Lake, where some people warned that fish populations had declined so much in recent years that they were now on the point of collapse. The problem was that very little real data existed on fishing effort or total catch. So we came up with the novel idea of working with a fishing family on the north shore of the lake for one year, during which we would document every single fish catch made in the family's arrow-shaped trap. This would provide information at a level that had never been collected before and enable us to establish a baseline to monitor future fisheries trends.

Coordinating this effort was Peng Bun Ngor, who had gotten his start at the Mekong River Commission two decades earlier and was now the fisheries science dean at the Royal University of Agriculture in Phnom Penh. Working with him were our Fulbright scholar Elizabeth Everest and Aaron Koning. Aaron joined our project to study fish reserves in the Tonle Sap Lake, which has one of the largest networks of fish reserves in the world. With fish reserves showing a marked increase in both diversity and biomass, especially of large-bodied fish, the Tonle Sap network, consisting of both government and community reserves, could be a game changer for fisheries management and conservation in the Mekong.

In northern Cambodia, our research team studied the impact of hydropower dams on fisheries in the 3S rivers—Sesan, Sekong, and Srepok. Originating in Laos and Vietnam, these Mekong tributaries provide as much as 25 percent of the Mekong River's flow and sediment load and support 40 percent of the fish in the system. In the last three decades, more than two dozen dams have been built in the 3S basin, though none in the Sekong River. Our researchers found a far greater fish diversity in the Sekong compared to the dammed rivers. These findings are important in the continued fight to keep the Sekong, which is now the target of several dam projects, flowing freely.

But we were also doing a lot of work that wasn't directly connected to fish. Sudeep, who is a limnologist, had overseen the development of a full limnology lab housed at the Inland Fisheries Research and Development Institute (IFReDI) in Phnom Penh. The lab, the first of its kind in Cambodia, would allow us to study a whole range of biological and chemical features of the Mekong River system.

Despite the pandemic, we continued to develop our outreach work. Our wonderful program manager, Chea Seila, initiated an educational campaign to combat plastic pollution in the Mekong, which would later result in several successful clean-up events. Chhut Chheana, our enthusiastic communications coordinator and jack-of-all-trades, visited schools around the country, as long as they remained open, distributing thousands of coloring books he had created with facts and figures about Mekong biodiversity.

Our project's outreach work was boosted by a dedicated group of "Young Eco Ambassadors," as well as a number of conservationists and activists who had been recognized as "Mekong Heroes." With the help of the communications team in the United States, we built a social media following of almost two hundred thousand people, mostly Cambodians, who could learn and take part in outreach events.

Our idea had always been to communicate the story of the Mekong to the broadest possible audience—in Cambodia, the United States, and around the world. National Geographic's philosophy of combining science and discovery to inspire and educate guided our thinking. From its start, Stefan, a longtime contributor to National Geographic, regularly published articles related to the project about the Mekong for National Geographic Online as part of a Wonders of the Mekong news series, shining a spotlight on both the positive efforts (people working to conserve fish, dolphins, and turtles) and the existential threats facing the Mekong River. Chheana, for his part, regularly produced video stories for Thmey Thmey, a popular Cambodian news outlet.

Eventually, the travel restrictions were lifted, and in early 2022 I was finally able to get back to Cambodia along with many of the other Americans working with us. The number of Cambodian researchers on the project had quadrupled since its inception, with scores of graduate students joining our team to earn their advanced degrees. Partnering with several universities in Cambodia and the United States, we had grown into the single largest research group of its kind based in the Mekong region.

In Cambodia, it was the height of the dry season, and temperatures were creeping up as team members split into groups and fanned out across the country. One group, led by Sudeep, set out to traverse the entirety of the Mekong River and the Tonle Sap system by boat. Spending more than a week zigzagging the waterways, the team measured water temperatures, algal concentrations, and oxygen and greenhouse gas levels in the river. It was the first time anyone had mapped the water quality and overall status of the Cambodian Mekong River system, from Vietnam to the Lao border.

Another group focused on the dai fisheries on the Tonle Sap River, trying to determine how many fish migrate down the river and into the Mekong annually. The traditional way to get any kind of answer had been to rely on catch data, which had a lot of limitations. But the team from one of our academic partners, the University of Washington, had found a way to use echosounders and other sophisticated technologies to monitor and visualize in real time the biomass of fish moving down the river.

Two separate groups had traveled to the country's north. One organized a citizen science effort to establish a network of local fishers to help us study fish catches over a large area, especially of endangered and understudied species. A second group, meanwhile, was tracking a previously unknown migration of the Mekong shad, a herring-shaped fish that has become endangered in the region. Meanwhile, on the Tonle Sap Lake, Aaron and his team continued monitoring fish reserves and recording all the fish caught by the family we were working with.

The energy and enthusiasm shown by everyone in the field was contagious. On our arrival in Cambodia, we had created a WhatsApp group that included all the project members and which we had dubbed Super Field Research Team. It was now overflowing with images of fish and all kinds of other discoveries that the team members were sharing from around the country.

The commitment and can-do spirit that I saw in the people working with us, especially the Cambodians, filled me with confidence that, with their help, we were making a positive difference for the Mekong. Despite the tremendous pressure the system was under, conserving it was not a lost cause. It was impossible to deny the life-giving properties of the river; you saw it everywhere you looked. It was there in the remaining flooded forest that continued to nourish the fish as it always had, and it was there in the river dolphin that gave birth to a baby calf. And if that life-giving essence could just be preserved, maybe it could be enough to protect the giant fish of the Mekong too.

But the river—and the fish—needed our help, in whatever way possible, and so while in Cambodia I organized a release into the wild of some critically endangered Mekong species, including a five-foot-long Mekong giant catfish with a story as unlikely as it was hopeful. It had begun thirteen years earlier when the catfish had been plucked out of the river along with thousands of other baby fish by a trader who was growing fish in her ponds at home, a common practice in Cambodia.

With time, the trader realized that five of the fingerlings she had collected were growing larger than all the other fish and were, in fact, Mekong giant catfishes. Still, the trader kept them until they grew too big for her ponds.

At this point, the trader could have sold the giant catfishes for meat and earned some decent profit. But she didn't. Knowing she was dealing with a critically endangered species, she instead contacted the local fisheries administration to see if it would take the fish off her hands, which it agreed to do since there was a perfect place to house those giant catfishes: the Freshwater Aquaculture Research and Development Center in the Bati village in southeastern Cambodia.

This research station has a series of large pools where endangered fish are reared. Since 2017 our project has partnered with the Fisheries Administration in running it. Every year we collect young fish from the Mekong and the Tonle Sap Rivers and place them in the ponds. The goal has been to one day release them back into the wild. We had already released a few individual fish here and there, but now I thought it was time to start doing it in a more organized way.

We decided to arrange for an event in the Tonle Sap Lake, where we would release more than a thousand river catfish, a species that was once a common food staple in Cambodia but has now also become endangered, along with dozens of juvenile giant barb, Cambodia's national fish, that we had reared at the Bati station. We would also release two Mekong giant catfishes: one a two-and-a-half-foot individual that we had grown at the station, and the other being one of the five catfishes we had received from the fish trader. The Cambodians on our team named this larger fish Samnang, the Cambodian word for "lucky."

For a reintroduction to succeed in the long term, the habitat must be healthy enough to support the fish. A study published in the journal *Conservation Biology* some years ago concluded that inadequately addressing the initial cause of decline in a fish population was the best predictor of "reintroduction failure." Another problem is introducing fish that have been bred in captivity—which is now commonly done with many large fish in the Mekong—into the wild, because a lack of genetic diversity often makes them unable to reproduce successfully. This is something we've seen over and over again all around the world.

But the fish we were releasing hadn't been bred at the Bati station. They had been collected as babies in the wild and retained their genetic integrity. As for releasing them into the Tonle Sap Lake, there was, for sure, an issue with the fishing pressure there. The likelihood of the fish getting caught, and killed, in the maze of nets around the lake was very high. But the habitat itself actually remained relatively intact, despite being disrupted by dams and drought, and, crucially, it also had the fishing reserves where we could release the fish. It made the Tonle Sap a good place to test a reintroduction.

The endangered fishes' two-day trip back to the wild began around 8:00 p.m. on a Wednesday, when two vans fitted with oxygenated water tanks departed the research center. One van carried hundreds of river cat-fish, along with juvenile giant barbs. In the other van were the two Mekong giant catfishes.

Traveling through the night, the vans arrived in the northwestern city of Siem Reap, just as the sun rose over nearby Angkor Wat. There, the fish were held in ponds and tanks at another research station, before being taken—along with hundreds more river catfish from that second station—the following morning by vans and then boats into the open waters of the Tonle Sap Lake, where we would release them into a government-operated fish reserve.

The fish had been tagged so that information about them could be gathered if they were recaptured, with fishers rewarded for returning the tags with information about the capture. On board a vessel from which the release would take place, we were joined by a large group of local officials and other people. A ceremony followed, during which Poum Sotha, the director-general of the Cambodian Fisheries Administration, urged fishers to help the research effort. I gave a short speech too, talking about the necessity to do more to save these iconic creatures that were found nowhere else in the world.

And then we began to release the fish. Among them was Samnang, the female Mekong giant catfish that had grown over five feet long but could still double in length, if given the time to continue to mature. As she disappeared into the Tonle Sap, I thought of the tag with which she had been marked as a sort of message in a bottle, and how messages in a bottle are often used to communicate distress. I expected the message that she was carrying to be found and listened to. But if no one ever received it, and Samnang was allowed to stay in the underwater world where she belonged, that was a good thing too.

A Wonder of the Mekong

The question I had started with—what is the world's largest freshwater fish?—was always meant to be a jumping-off point for a larger inquiry into the health and status of the world's river giants and their habitats. But I still wanted to solve the mystery. The fact that I didn't find the answer as quickly as I had expected only motivated me to continue the search.

There were several contenders for the heavyweight title. One, the Chinese paddlefish, sadly had disappeared during the early stages of my research. It may, in fact, have been the world's largest freshwater fish; the story published in *Science,* with a report of an individual twenty-three feet long, suggested so. But there was, to my knowledge, no verified account of a record-breaking weight of a paddlefish.

Mostly, the habitats for giant fish were deteriorating. But there was one megafish, the wels catfish, that was actually thriving in altered conditions, and growing enormous. While many giant fish did not reach their maximum age (and size), the wels was an exception. Maybe the fact that populations of this species were spreading, and the fish living longer, meant that wels catfish would grow bigger than any other fish.

I had never dismissed the arapaima as a contender. One day I got an email from Donald Stewart, the arapaima researcher. He and his colleague Cynthia Watson had been studying arapaima in the Essequibo River in Guyana, estimating age, growth, and mortality of this population, which had recovered dramatically in recent years after being severely overfished. Analyzing growth ring deposition on the arapaima's scales, the two had arrived at a length-mass relationship that showed Guyanese arapaimas were much heavier than those from central Brazil. An arapaima measuring three meters and weighing 540 pounds had recently been caught in Brazil, Stewart said. The research in Guyana suggested that an arapaima of similar length there would weigh over 700 pounds.

For a verified weight of an actual catch, the Mekong giant catfish captured by the Thai fishers in 2005 was still the record holder at 646 pounds. But I thought there were probably bigger freshwater fishes out there, and there was one species specifically that I concluded could grow bigger than that: the giant freshwater stingray. I strongly suspected that the stingray I had handled on the Mae Klong River, which we had not been able to weigh because we couldn't get a scale there in time, was bigger.

Dr. Ning had also measured some very big rays, one with a disc width of over seven feet. At that size, every inch of width in the fish can result in dozens of extra pounds. As with the formula that Stewart produced for arapaimas, it was possible to calculate weight estimates for stingrays from their measurements. The problem was that we hadn't been able to weigh many rays, for fear of harming them, so our sample size was small. Of the twenty or so stingrays we had tagged for our movement study in the Mae Klong, only five had been weighed, making our estimates of maximum weight prone to error. Even so, we knew that a ray of six feet across weighed about four hundred pounds and at seven feet across it would likely approach a record-breaking weight.

The chemical spill on the Mae Klong appeared to have killed off the largest stingrays in that river. But the species was found elsewhere, too, and not just in mainland Southeast Asia. Beginning in 2016, it had been documented on the Indonesian island of Sumatra, with reports of a dozen large ray catches in local rivers. The reports were compiled by a biologist at Sumatra's Sriwijaya University named Muhammad Iqbal, and in late 2019 Stefan went to see him there.

From the city of Palembang, they traveled in a speedboat for several hours down the choppy Musi River, where most of the catches had been reported, until reaching the small fishing village of Bungin. There, a fisherman named Kamar said he had caught a giant stingray in a net in the open estuary where the Musi and other rivers emptied into the South China Sea. It had taken fifteen men to lift the ray out of the water and onto land, Kamar said. Borrowing a large scale from a local storage facility, the fishers had weighed the ray, with the scale showing that it weighed almost 400 kilos, or nearly 880 pounds.

It was a story not unlike other stories I had heard from fishers catching enormous stingrays. What lent Kamar's account extra credence was that not only had the fish been weighed, but Kamar had also saved the ray's skin after it was cleaned out and sold for meat. The skin had since shriveled, but its size suggested a ray with a width of close to eight feet, big enough to beat the record.

But the catch also raised questions. It was made in an estuary, in brackish water. There was no doubt the ray was the *Urogymnus polylepis,* based on its shape and markings, but it could belong to a population that was not strictly confined to fresh water. It added another layer to the inquiry. As with the white sturgeon of North America, there seemed to be ocean-going populations of the giant stingrays, as well as groups restricted to fresh water.

There was one place where I was certain the stingrays spent all their lives in fresh water, and that was the Mekong River in the upper reaches of Cambodia, as it descends from neighboring Laos, flowing languidly past sandbanks and islands covered in seasonally flooded forest. In this stretch of river, about a hundred miles in length, up to two hundred billion fish are spawned each year. From a biodiversity standpoint, there's no more important section of the river, with deep pools reaching down 260 feet that are believed to be dry-season refuges for Mekong megafish species, including giant catfishes, giant barb, and giant stingrays.

This area had been a focus of our Wonders of the Mekong project from the start. Like much of the river system, it was coming under increasing pressure from overfishing, different types of pollution, and droughts exacerbated by climate change. Dam operators upstream, primarily in China but also in Laos, withheld water for their own economic gain, causing water shortages. And there was a long-standing proposal to build Cambodia's first hydropower plant on the main stem of the Mekong, in the exact place where the deep pools are found.

But the remoteness of the area made it difficult to study. This was true for the environment above water. Finding out what was happening below the surface—not to mention all the way down to the deepest part of the Mekong—was a lot more problematic. But it was something I had been wanting to pursue for a very long time. So in early 2022 I enlisted the help of the Wonders of the Mekong team and two National Geographic Explorers: Kenny Broad, an environmental anthropologist and world-class diver, and Kakani Katija, a deep-sea explorer who runs the Bioinspiration Lab at the Monterey Bay Aquarium Research Institute. We began, for the first time, to explore the Mekong's deep pools.

For several days we dove down to this never-before-observed environment, using unmanned submersibles equipped with lights and cameras: drop cameras suspended on long cables, and baited video cameras as our eyes and ears. In the turbid deep of the Mekong, the visibility was almost zero, and the remote-operated vehicles stirred up a ton of silt when they landed on the bottom or drove along it. And yet, seated in the boat atop the river surface, looking at the computer screen showing the dark, mysterious habitat below us, I felt, for the first time, intimately connected to this hidden world that I had studied for so long.

Despite the visibility challenges, we were able to capture fleeting images of fish. More than two hundred feet down, a school of catfish swam by the cameras. They were *Pangasius macronema,* one of the migratory species I'd studied at the Khone Falls all those years ago. These fish had once formed the basis of the communal fishery across the border in Laos, which no longer existed because the fishes' migratory path had been blocked by the building of the Don Sahong Dam. I had wondered what happened to these fish. Here I was given a lead to pursue.

I knew the giant stingrays were down there, too. But the odds of catching one on camera were next to nil. The best source of information about the rays was, once again, going to be the fishers who were out on the river every day pulling up fish. As part of the project, we had conducted community and market surveys to gather information about aquatic biodiversity in the area. In collaboration with Cambodia's Fisheries Administration, we had also begun establishing a network of fishers who agreed to report catches of giant and endangered fish.

In the past, fishers had been reluctant to do so. Because of the challenges involved in catching these animals, and also because they were not considered a good food fish, the giant stingrays were generally not targeted by fishers. But they could be captured by accident on baited hooks or caught using trotlines. Such incidents were rarely reported. It wasn't illegal in Cambodia to fish for the giant freshwater stingray, which is listed as endangered by the IUCN, but many fishers seemed to believe that catching huge rays was still something that could land them in trouble. If and when a stingray was caught, it was usually killed and butchered quickly, erasing all evidence of its existence.

Because of our outreach efforts, however, the fishers had become more willing to share their experiences with the rays. When I went to talk to them now, they regaled me with stories of catching huge stingrays. One older man seemed to take particular pleasure in recounting every detail of his several catches of record-sized rays. His name was Suong Seun, and he told me about three catches of stingrays weighing more than four hundred kilos, each of which would have been the largest freshwater fish in the world. Unfortunately, he said, he had not had a camera on hand with which to take photos of the fish.

I sat there listening to his stories, and the stories of his fellow fishers. I asked for details about how and where each stingray had been caught, reframing my questions in what seemed like a never-ending search for clues that could either confirm or invalidate the testimonies. But it was futile. I was never going to get an answer of absolute certainty based on the oral accounts of fishers. It wasn't that I didn't believe them—in fact, I was certain their claims of pulling giant rays out of the river were largely true—but without confirmed weights or any kind of scientific verification, their stories remained just that: stories.

I began to accept that my search for the world's largest freshwater fish had come up short. To be clear, I didn't see this as a failure. So much had been gained and learned about megafish by asking the question. But it would be a lie to say I wasn't disappointed. The point of asking a question, after all, is to find an answer. Besides, the fact that neither I nor anyone else had been able to confirm a record-breaking fish suggested something ominous: perhaps the record had not been broken because of the global decline of megafish populations. Maybe the true giants were permanently in our past, and there were no longer any fish big enough to break the record.

A few days had passed since our deep-pool expedition, and I was back at home in Reno. I woke up one morning to a text message from Seila, our project manager in Cambodia. The message said that fishers had caught a giant stingray in our Mekong study area. One of them had contacted Chhut Chheana, our media man who had helped build the fishers network, asking if our team could come to a remote island in the Mekong to help release it.

It was around 9:00 p.m. and night had long since fallen when our team, which included Seila, Chheana, and our Fulbright scholar Elizabeth Everest, set off by van from Phnom Penh for what turned out to be a six-hour journey through driving rain. When they arrived at the Mekong River downstream from the town of Stung Treng, it was still dark, and so the team members decided they'd sleep in the van for a few hours before moving on to the island once dawn broke.

On the island, they were met by the fisher who said he had caught the stingray accidentally on a fishing line the previous night. Apparently the ray had swallowed a smaller fish that in turn must have eaten one of his baited hooks. With the help of his brothers, the fisher had managed to bring the stingray to the shore, where it had been tied up with a rope and kept in the water.

More than twenty-four hours had passed since the capture, but as the fishers pulled the stingray out of the water, it appeared to Seila and the others to be in good health despite the stress of being caught. It was a very large female, but it soon became clear that it was not of record-breaking size. It measured 3.93 meters, or almost 13 feet, from the snout to the tip of its tail. This was about a half foot shorter than the first ray I ever saw, in southeastern Cambodia, twenty years earlier.

To weigh the ray, Seila had brought three 100-kilo scales, which were placed next to each other with a plywood board on top of them. When the ray, which otherwise had been kept in the water since the capture, was lifted on to the board, the scales showed a total weight, adjusted for the weight of the board and tarp, of 180 kilos, or almost 400 pounds.

Although it was not a record-breaking fish, its size was still very impressive, even to the many villagers who had congregated on the shore. Some of them had never seen a giant stingray before. The chief of the local commune, a man named Long Tha, crouched down next to a young boy. "This is a giant stingray," the chief told the boy. "When you grow up, you should protect it."

Our project compensated the fisher for agreeing to release the stingray. This was similar to the buy-back approach implemented during my early field studies on the Mekong. It was a way to incentivize fishers to call us when they caught stingrays or other rare fish, rather than kill the fish and sell the meat. But it was important not to pay more than market value, since that might cause fishers to start targeting stingrays for profit. We knew the market rate for stingray meat was three dollars per kilo, so that's what we paid.

Having taken the measurements and photos and video of the stingray, it was time to release it. As the ray was pushed into the water, the community members cheered. "Bye, bye," someone shouted. At first, the stingray moved along the surface, raising its nose to test the waters, perhaps not believing it had been released back into the river. It even picked up its tail and splashed the surface, as if to say goodbye, before eventually slipping into the darkness below.

It was the middle of June and the rainy season had officially begun in Cambodia. I had been under the impression that stingrays were almost always caught in the dry season (and primarily from December to April), but since that first ray was captured—in early May, at the tail end of the dry season—fishers had caught two more stingrays, which our team had helped them release. The rays were not as big as the 400-pound individual, but the captures suggested that giant stingrays were present in the area even after the rains had arrived.

At this time, we had a team of scientists from our California-based partner organization FISHBIO in the region to set up an array of underwater receivers in the Mekong, from the Lao border down to Kratie and into the 3S Rivers. Our plan was to implant acoustic tags in up to two hundred fish, which would be tracked by the receivers to learn more about the fishes' migratory patterns and behavior. It was the most extensive telemetry project ever implemented in Cambodia.

A few days before the team was scheduled to wrap up its work, I sent an email to the Wonders of the Mekong group asking if there had been any more reports of stingrays. I reminded people that we had the FISHBIO scientists in the field who would be able to tag any fish. Only a few hours had passed when I received a WhatsApp message from Chheana. It said that a fisher had called him to say that another, "bigger" stingray had been caught.

It was now close to 8:00 a.m. in Cambodia, and I immediately called Chheana, who had just finished his morning exercise routine. He gave me the basic details of what the fisher, who lived on the same island where the other rays had been caught, had told him. The stingray had been captured overnight on a fishing line with baited hooks, much in the same way that the other rays had been caught. It was still alive and held at the river's edge. I asked Chheana what he meant by "bigger" stingray. He said the fisherman had told him that it was "much bigger than the others," meaning the 400-pound ray.

I figured there must be a reason for the fisherman to make such a claim. At the same time, I didn't want to make too much of it. I knew that until a fish was weighed, any claim about its size, and especially when it was based on third-hand information, was meaningless. I had been through too many false stories for that. Besides, my instant reaction focused on the opportunity we had to tag and track—for the first time ever—a giant freshwater stingray in an area believed to be a stingray hot spot and a possible spawning site.

But we had to come up with a plan, and fast. Coordinating this kind of mission required a lot of logistics and details to be worked out. And so I spent the next two hours on the phone with the various team members, and on group calls that included Dana Lee and Jack Eschenroeder from FISHBIO. Those guys had been up in the 3S River basin to install receivers, but were happy to drop what they were doing and get themselves down to the Mekong in order to (hopefully) carry out the tagging of the stingray.

It was now late morning in Cambodia. We agreed that the goal should be to tag and release the stingray that same day, and before dark—that is, before 6:00 p.m. We did not want the stingray to be held captive for more than twenty-four hours. This didn't give us a lot of time. Thankfully, Seila and Chheana were able to quickly organize their departure from Phnom Penh, with the FISHBIO team, which also included Doug Demko and Sinsamout Ounboundisane, as well as Seat Lykheang, the team leader for our project's Young Eco Ambassador program, planning to meet them on the banks of the Mekong where a ferry would take them to the island.

I continued to stay in contact with Seila and Chheana during their van journey up north, sorting out the details of who would do what. We also had to liaise with our partners in the Cambodian Fisheries Administration, who were going to be involved with the tagging. There was no way of telling how things would turn out. But I knew we were well prepared and well equipped for the job. We had the trust of the fishing community. In fact, the fisher who had caught the ray had told Chheana that he was keeping the fish in an undisclosed location and was not going to show it to anyone before our team arrived. And so I went to bed, around 1:00 a.m., secure in the knowledge that if anyone could pull off this kind of mission, it was going to be our team.

I'd been asleep for a little more than an hour when my phone buzzed. It was a text message from Seila saying the group had arrived at the ferry, where they had met up with the FISHBIO folks, and they were now heading over to the island. So far, so good, I thought, and went back to sleep. Then, less than two hours later, at 4:16 a.m., I was awakened by yet another text message. This time Seila wrote: "stingray successfully tagged and released."

My immediate reaction was a mixture of relief and excitement. The stingray had been safely returned to the river where it belonged, and our team had been able to successfully tag it. We would now be able to follow its movements. This was a huge conservation win, and it had the potential to be a game-changer in our quest to learn more about these animals and to protect them.

And so when I went back to sleep this time, it was with a sense of satisfaction about what we had been able to accomplish, as well as excitement about the challenging work that still lay ahead. And this time I did not wake up when my phone once again buzzed, at 5:03 a.m. It was not until after 6:00 a.m., the time I usually got up, that I finally saw that I had received one more message from Seila. This one read:

TOTAL WEIGHT IS 300KG

At 661 pounds, it was the largest freshwater fish ever recorded.

Guinness World Records

Record-Breaking Ray Confirmed As World's Largest Freshwater Fish

By Adam Millward • Published 24 June 2022

Fishermen in the Stung Treng region of northern Cambodia made a historic catch on 13 June 2022 when they hauled in the largest freshwater fish to ever be officially recorded.

The female giant freshwater stingray (*Urogymnus polylepis*) weighed approximately 300 kg (661 lb)—about the same as a typical grizzly bear—and had a total length of 3.98 m (13 ft) with tail included, making her longer than a pickup truck!

The phenomenal fish, also known as a whipray, spanned 2.2 m (7 ft 2 in), meaning that if she were placed on a ping-pong table, her outer "wings" would overhang each side by a foot (30 cm).

The record-breaking ray, which is an endangered species, was named Boramy ("full moon" in the Khmer language), owing to both her rounded disc-like shape and the early-evening time of her release.

Although discovered by locals, the measurements were taken by a team of international ichthyological experts working as part of the US-Cambodian "Wonders of the Mekong" project, in collaboration with the Cambodian Fisheries Administration.

[According to] giant fish expert Dr Zeb Hogan, who presents the TV series *Monster Fish* but also serves as director of the "Wonders of the Mekong" initiative: "This is an absolutely astonishing discovery and justifies efforts to better understand the mysteries surrounding this species and the incredible stretch of river where it lives."

Source: Millward, Adam (managing editor), Guinness World Records.
See full article at www.guinnessworldrecords.com/news/2022/6/record
-breaking-ray-confirmed-as-worlds-largest-freshwater-fish-708416.

Acknowledgments

The stories in this book come from many years of research, travel, and friendships and the support of hundreds of people who have provided their time, kindness, and advice. I especially would like to acknowledge my coauthor and friend Stefan Lovgren and our agent Will Francis at Janklow & Nesbit. This book benefited from the guidance and assistance of Sudeep Chandra, Peter Moyle, Bernie May, Brant Allen, Barbara Barnes, Michele Thieme, Steve Weiss, Carol Bender, Silvia Hillyer, Niek van Zalinge, Cheryl Zook, Rebecca Martin, Anna Rathman, John Francis, Rob Lee, Kathryn Keane, Seth de Matties, Dean "Johno" Johnson, Rob Taylor, Geoff Daniels, Jatuporn Athasopa, Paulo Velozo, Drew Pulley, Geoff Luck, Abigail Pilgrim, Jake Klim, Chelsea Mose, Jeff Thompson, Jack Hayes, Jane Tors, Mike Wolterbeek, Elizabeth Everest, Mat Sinclair, Susan and Gary Clemons, Brian Eyler, Suthep Kritsanavarin, Chainarong Sretthachau, Ian Baird, Thach Phanara, Em Samy, Srey Pheap, Mout Sitha (and Sokha, Phea, and Phors), Praichon Intanujit, Seth Mydans, Sarah Null, Jamie Voyles, Bonnie Teglas, Jon Stout, Teresa Campbell, George Naughton, Heng Kong, Peter Starr, Marco Barbieri, Curtis Vickers, Ricky Verrett, Allyse Ferrara, Dave Morgan, Dennis Scarnecchia, Solomon David, Dan Dougherty, Kirk Kirkland, Richard Kissling, Don Stewart, Julien Cucherousset, Frédéric Santoul, Trip Jennings, Andy Maser, Flavio Lima, Vinay Badola, Will Crampton, Wei Qiwei, Jake Vander Zanden, Olaf Jensen, Pete Rand, David Thawley, Mimi Kessler, David Gilroy, Erdenebat Manchin, Jeff Liebert, Dan Vermillion, Bazarsad Chimed-Ochir, Erdenebat Eldev, Tsoqtsaikhan Purev, Sarantsetseq Borchuluun, the Mongolian science camp team (Amaraa, Moogi, Ganzorig), Charlie Conn, Steve Correia, Grady Glyer, Sok Long, Nantarika Chansue, Rick Humphreys, Wuttichai Khuensuwan, Dion Wedd, Robbie Bridgman, Olfi Pesca, Günther Grill, Herman Wanningen, John Zablocki, Johannes Schöffmann, Anders Poulsen, So Nam, Ian Harrison, Will Darwall, Danielle Williams, Velibor Ivanovic, Stuart Gillham, Fengzhi He, Abigail Lynch, Sonja Jähnig, Lisiane Hahn, Eric Baran, Steven Cooke, Julian Olden, Caroline Pollock, Lee Baumgartner, J. David Allan, Robin Abell, David Philipp, Julie Claussen, Peter-John Meynell, Alvin Lopez, Aaron Koning, Grant Rader, Erin Buxton, Duane de Freitas, Ashley Holland, Rene Henry, James Simmons, Pippa Newell, Mary Melnyk, Ngor Peng Bun, Chea Seila, Saray Samadee, Chhut Chheana, Rachel Nuwer, Jason Bittel, Gene Helfman, Kirk Winemiller, Craig Paukert, Richard Stone, David Braun, Julia Louie, Steve Pickard, Melanie Stiassny, Katherine McCall, Sander Van Der Lin, Rajeev Raghavan, Gordon Reid, Sandra Postel, Terry Garcia, Alex Moen, Sarah Fowler, Borja Heredia, William "Bill" Perrin, Ronald Bruch, Michael Mars, Chris Barlow, Tierney Thys, Kenny Broad, Kakani Katija, Enric Sala, Sylvia Earle, Michael Fay, Elena Guarinello, Ibadet Dervishaj, Patrick Truby, Alan Parente, JoAnne Banducci, Jinni Fontana, Sara and Matt Hendricksen, Chheng Phen, Thay Somony, Nao Thouk, Chavalit Vidthayanon, Noppakwan Intaphan, Erin Loury, Shaara Ainsley, Dee Thao, Elizabeth Ramsey, Jack Eschenroeder, Dana Lee, Thom Benson,

Jessica Cohen, Tim and Kathy Hogan, Zack and Dale Hogan, and Margaret Shelton. I deeply appreciate the love, patience, and support of my wife Abby Hogan and the energy and curiosity of my son Wes.

I also would like to acknowledge the organizations that directly or indirectly supported this book, including the Flinn Foundation, the University of Arizona Undergraduate Biology Research Program, the USGS Arizona Cooperative Fish and Wildlife Research Unit, the Fulbright US Student Program, the National Science Foundation, the University of California, Davis, Graduate Group in Ecology, the Mekong River Commission, the IUCN Mekong Wetlands Biodiversity Conservation and Sustainable Use Programme, the Wildlife Conservation Society, the World Wildlife Fund, the National Geographic Society, the University of Nevada, Reno, College of Science, the University of Nevada Press, the Nevada Discovery Museum, and the US Agency for International Development. I would like to thank the people living in communities on the rivers and lakes around the world, for their hospitality and for teaching me about the important connection between water, fish, and people.

—Zeb Hogan

In 2006, I was asked by National Geographic News to go to Cambodia to do a story about Zeb Hogan and his conservation work. Zeb had been named an emerging explorer by National Geographic. It would be the start of many years of adventuring up and down the world's great waterways in search of some very big fish. I am deeply grateful to Zeb for taking me on this journey of wonder and entrusting me to write this book with him.

I'd like to thank the people at National Geographic who have supported my text and video stories over the years, beginning with David Braun, Sean Markey, Mark Bauman, Jeff Hertrick, and Ted Chamberlain, and continuing with Brian Clark Howard, Christine Dell'Amore, Rachael Bale, Lori Cuthbert, and Rob Kunzig.

Since I didn't know much about fish and freshwater issues when I began writing about this, I am indebted to all the conservationists, fishers, rangers, ecologists, anthropologists, and many others (-ists and otherwise) who have helped me understand the complexities of these subjects. A special thanks to the more than one hundred experts in their fields whom I've interviewed specifically for this book. I feel a lot more confident talking about potamodromous fish at cocktail parties now.

Many people helped us bring the book from idea to finished manuscript, starting with our agent Will Francis at Janklow & Nesbit, who did a super job with the book proposal. It's been a pleasure working with the folks at University of Nevada Press, including JoAnne Banducci, Jinni Fontana, and (early on) Sara Hendricksen. A big thank you to Curtis Vickers for making all the right cuts and changes, and to John Mulvihill for fixing everything on the home stretch.

I am appreciative of the early financial support of our book from the University of Nevada, Reno, College of Science and its former dean, Jeff Thompson. Through the university, I've been part of the Wonders of the Mekong project, where it's been a privilege to work with people like Sudeep Chandra, Bonnie Teglas, Erin Loury, Aaron Koning, Dee Thao, Elizabeth Everest, Elizabeth Ramsay, and all of our amazing

Cambodian colleagues, including Chhut Chheana, Chei Seila, Thach Phanara, Ngor Peng Bun, Saray Samadee, and Lykheang Seat.

Writing any book is a time-consuming process, especially one requiring as much research and travel as this project. It would have been impossible for me to pull this off without the patient love and understanding of my wonderful wife Shannon and our three awesome boys, Lukas, Elliot, and Aron.

Finally, I'd like to express my admiration for all the people around the world who are tirelessly working to protect the rivers, lakes, and wetlands that sustain so much of life on Earth. I can think of no more important work than what you do. Thank you!

—Stefan Lovgren

Index

About the Authors

ZEB HOGAN is an associate professor at the University of Nevada, Reno, a National Geographic Society Explorer, and the United Nations Convention on Migratory Species Scientific Councilor for Fish. He has a doctoral degree in ecology, was a visiting Fulbright scholar at the Environmental Risk Assessment Program at Thailand's Chiang Mai University, and has served as a World Wildlife Fund senior freshwater fellow. Hogan also hosts the Nat Geo Wild television series *Monster Fish*. Zeb's research focuses on migratory fish ecology, fisheries management, and endangered species issues.

STEFAN LOVGREN is an award-winning journalist and best-selling author with more than twenty-five years of reporting experience from around the world. He has been a regular contributor to National Geographic's various media platforms since 2003, writing about a wide range of environmental issues with a particular focus on fish and freshwater subjects. He produces print and video media content for the University of Nevada, Reno's Wonders of the Mekong project. Lovgren holds a master's degree in international affairs and journalism from Columbia University in New York. He currently lives in Las Vegas, Nevada.